Three principal figures emerge from this case. Combined, their lives explain how and why La Cosa Nostra, the once-monolithic and seemingly impenetrable criminal organization, has come apart at the seams.

There is Joseph "Skinny Joey" Merlino, a second-generation South Philadelphia wiseguy, an MTV mobster whose style and flair were overshadowed only by his insatiable desire for money.

There is Ralph Natale, the sixty-going-on-thirty Mafia don living thirty years in the past, who became the first sitting mob boss ever to co-operate with the Feds.

And there is Ron Previte himself: an underworld mercenary—a cold, calculating, one-of-a-kind mobster whose only loyalty was to the man staring back at him in the mirror.

One of them is

THE LAST GANGSTER

You decide.

Books by George Anastasia

THE LAST GANGSTER
THE SUMMER WIND
THE GOODFELLA TAPES
BLOOD AND HONOR
THE BIG HUSTLE
MOBFATHER

the last GANGSTER

From Cop to Wiseguy to FBI Informant:
Big Ron Previte and the Fall of the American Mob

George Anastasia

ReganBooks

AVON BOOKS
An Imprint of HarperCollinsPublishers

Photo insert credits: pages 1, 6, 7 (bottom), 8, 13 (top) Ron Previte; pages 2, 3, 4, 14, 15 (top) Federal Bureau of Investigation; page 5 Andrea Mihalik (courtesy *Philadelphia Daily News*); page 7 (top) *Philadelphia Daily News* archive; page 9 David Maialetti (courtesy *Philadelphia Daily News*); page 10 (top) Jim MacMillan (courtesy *Philadelphia Daily News*); pages 10 (bottom), 11, 12 (top & center), 13 (bottom) Brad Nau; page 12 (bottom) Ron Cortes (courtesy *Philadelphia Inquirer*); page 15 (bottom) G. W. Miller III (courtesy *Philadelphia Daily News*); page 16 Alejandro Alvarez (courtesy *Philadelphia Daily News*)

AVON BOOKS
An Imprint of HarperCollins*Publishers*
10 East 53rd Street
New York, New York 10022-5299

First Avon Books/ReganBooks paperback printing: March 2005
First ReganBooks hardcover printing: March 2004

The ReganBooks hardcover edition contained the following Library of Congress Cataloging-in-Publication Data:

Anastasia, George.
    The last gangster : from cop to wiseguy to FBI informant : big Ron Previte and the fall of the American mob / George Anastasia.—1st ed.
        p. cm.
    1. Previte, Ron. 2. Mafia—Pennsylvania—Philadelphia Metropolitan Area—Biography. 3. Mafia—Pennsylvania—Philadelphia Metropolitan Area—Case studies. 4. Organized crime—Pennsylvania—Philadelphia—History. I. Title.

HV6452.P4M34258 2004
364.1'06'0974811—dc22   2003068750

For Angela, Michelle, and Nina.
Luck be a lady.

PROLOGUE

They sat in a booth in a South Philadelphia diner in the fall of 1998 discussing life in the underworld—the scams, the shakedowns, the money made, and the money lost. The key, Johnny Ciancaglini said, was to keep a low profile.

"Just make yourself a little scarce. That's what I do. I mean, I'm there if anybody needs me, but I try to be as scarce as I can. I mean, you'll knock twenty, thirty percent of your problems out."

Ron Previte nodded in agreement.

They had been talking for about twenty minutes, ironing out a problem that went back five years, to the day when Previte and another wiseguy had taken ten grand from a bookmaker who was with Ciancaglini. "Johnny Chang," as Ciancaglini was known in certain circles, was in jail at the time, doing seven years on an extortion rap. The bookmaker, and the action he had on the street, had been Chang's sole source of income, and the shakedown had a direct impact on him.

"After he gave you the ten, he folded up," Chang said. "My income was dead out. . . . He folded up and went to Florida. . . . I mean, that hurt me big time because that let all my income go."

They were drinking coffee. Sitting in a booth in the Oregon Diner, a popular no-frills veal-parmigiana-with-soup-and-salad kind of joint that has been operating on the corner of Third Street and Oregon Avenue for as long as anyone can remember.

John Ciancaglini was a South Philly guy. He was forty-four at the time, and was making a comeback after his stint in prison. Solidly built, with thick black hair and rugged good looks, he had both the pedigree and the background to be a player. His father, Joseph "Chickie" Ciancaglini, was a mob capo doing thirty-five years on a racketeering conviction. Johnny also had two brothers who had given blood for the organization. One was dead. The other was crippled.

Considered both street-smart and intelligent, John Ciancaglini knew how to make money. More important, he knew how to survive.

So did the guy sitting across from him in the Oregon Diner. "Big Ron" Previte, a six-foot, three-hundred-pound wiseguy, could grind it out with the best of them—had been doing it for years. An ex-cop and one-time casino security worker, Previte, who was fifty-five, had made several million dollars even before he was formally initiated into the organization. Drugs, prostitution, extortion, gambling, loansharking: you name it, he did it. He knew the business of the underworld inside out.

In fact, there were many who believed that the only reason Previte was still alive was because of his uncanny ability to come up with cash. And anyone who looked at things realistically would have agreed. The fact of the matter was that on November 2, 1998—the day he was sitting in the Oregon Diner drinking coffee with John Ciancaglini—Ron Previte should have been dead. Or in jail. Logic and common sense told you as much.

But logic and common sense were in very short supply in the Philadelphia underworld at that time.

Previte had been a soldier under mob boss John Stanfa during the bloody war that rocked the Philadelphia underworld in the early 1990s. The Stanfa organization clashed with a group of young mobsters headed by Joseph "Skinny Joey" Merlino and Michael "Mikey Chang" Ciancaglini, John Ciancaglini's youngest brother.

Adding another twist to the story was the fact that Stanfa's underboss was Joseph Ciancaglini Jr. Joey Chang was the middle brother in the family. From 1992 to 1994 the streets of South Philadelphia were a battleground, as teams of hit men stalked one another in a struggle for control of the organization.

The battle lines were both cultural and generational. Stanfa, who was in his fifties, was an old-school, Sicilian-born mob boss who was more at home on the streets of Palermo than in South Philadelphia. Merlino and Mike Ciancaglini, both in their early thirties, were flamboyant young wiseguys, street-corner gangsters. The war ended with a half-dozen people dead and with Stanfa and most of his top associates in jail.

By the time Previte met with Johnny Chang at the Oregon Diner in 1998, though, that was all in the past. Now it was time to mend fences. The sit-down was arranged to smooth over any hard feelings Johnny Chang still harbored after the shakedown of Chang's bookie.

"Bygones were bygones," Merlino had said when he set up the meeting.

Previte, who had been with Stanfa during the war, was now with Skinny Joey.

Or so everyone thought. Through it all, in fact, Ron Previte had been with the FBI. A confidential informant of the highest order, he was wearing a body wire as he sat in the diner that afternoon. The conversation with Johnny Chang was one of more than four hundred he recorded during a two-year period that began in February 1997 and ended in June 1999.

There has never been a gangster like Ron Previte. This is his story, told in part through his own words and in part through the voluminous law enforcement files that outline the government's multipronged but flawed case against Joey Merlino and the Philadelphia mob.

Three principal figures emerge from that case.

Each is a unique example of the American underworld at the turn of the century. Combined, their lives explain how and why La Cosa Nostra, the once monolithic and seemingly impenetrable criminal organization, has come apart at the seams.

One of them is the last gangster.

You decide.

There is Joseph "Skinny Joey" Merlino, a second-generation South Philadelphia wiseguy, an MTV mobster whose style and flair were overshadowed only by his insatiable desire for money. Handsome, charismatic, and deadly, Merlino gave a face and a personality to a mob family that for years had lived in the shadows of its bigger, bolder, and more infamous New York brethren.

There is Ralph Natale, a sixty-going-on-thirty Mafia don who was living twenty years in the past as the underworld in which he operated rushed toward the twenty-first century. Constantly boasting that he was a "man's man," Natale crumbled, like so many of the rats and informants he claimed to disdain, when he was faced with a narcotics conviction that could have sent him back to prison for the rest of his life. He became the first sitting American mob boss to cooperate with the feds, a distinction that adds to the ever-expanding underworld infamy of the Philadelphia crime family.

Finally, there is Previte, an underworld mercenary, a one-of-a-kind mobster whose only loyalty was to the person he saw staring back at him in the mirror each morning. Cold, calculating, and completely amoral when it came to making money, he put his life on the line each day he worked for the FBI.

Unlike Natale, who claimed to have had an epiphany that led him to renounce La Cosa Nostra after being charged with the narcotics rap, Previte doesn't claim that God, morality, or a sense of righteousness led him to do what he did.

It was simply a question of survival.

That and the realization that the Mafia—the mob, La Cosa Nostra—was not what it was cracked up to be. Previte had aspired to be a made member of the organization. He saw it as the pinnacle, the top of his profession. When he got there, though, he was disillusioned. Honor and loyalty had been replaced by greed and treachery. The sense of family that supposedly governed the actions of men of honor was gone. Instead, it was every man for himself.

"It was over," Previte says. "You'd have to be Ray Charles not to see it."

Previte cut his deal with the government because it was the smart, the sensible, the logical thing to do. He makes no apology for it.

At the diner sit-down, Johnny Chang told Previte he was offended by the shakedown. He took it personally.

"It's a little disrespectful toward me," he said. "For a guy . . . who's sitting in [jail] doing his fucking bit, to have somebody approach somebody that's his only income . . . You understand? Hey, listen, I'm not making the money an issue. The money's not an issue. . . . You know what I mean, Ronnie?"

Previte nodded. He said he understood. But he wanted Chang to understand as well. It was business. His boss at the time, Stanfa, had told him to get the money.

"If he tells me to do something, do I do it or don't I do it?" Previte asked, already knowing the answer.

"Well, yeah, you do," Ciancaglini conceded.

"It's nothing personal," Previte said. "We shook him down because the boss tells us to shake him down, Johnny."

"I'm just saying that what happened wasn't right," Chang said.

"I'm sure of that," Previte said.

"The only people that got screwed out of this whole thing is my family. I mean, look at it all the way around. My brother Joey can't fucking walk and chew gum at the same time. Michael's in a box."

Ciancaglini then apologized for any problems he had caused Previte. Previte, in turn, apologized for anything he might have said or done that offended Ciancaglini. Two mobsters, sitting around making peace and planning a better future.

Neither one was particularly crazy about the leadership at the top of the organization, but each was smart enough to realize there wasn't much he could do about it.

"You know what you got to do, Ronnie?" Ciancaglini said. "Be the man that you are, the man that you are. Just make yourself a little scarce. . . ."

Three years later, Ciancaglini was sitting at a defense table in U.S. District Court in Philadelphia as Previte, from the witness stand, buried the Philadelphia mob.

1

On the day he was shot, Joe Ciancaglini arrived for work at 5:54:46 A.M.

We know this down to the second because the FBI surveillance camera that was mounted on the telephone pole across the street from Ciancaglini's business establishment—a hole-in-the-wall coffee shop called the Warfield Breakfast & Luncheon Express—was taping that morning.

Dark-haired and handsome, the thirty-four-year-old Ciancaglini had been in the FBI's sights for several months, ever since he was named underboss of the Philadelphia mob. It was a move that John Stanfa, the city's mob boss, hoped would bridge a growing gap between the established organization and a younger faction that had been balking at Stanfa's rule.

Ciancaglini had the background, and the bloodlines, for the job. His father, Joseph "Chickie" Ciancaglini Sr., was doing heavy time on a racketeering conviction. His older brother, John, was serving seven years in an extortion case. His younger brother, Michael, was one of the leaders of the faction that was giving Stanfa trouble.

The prosecutions, convictions, and factionalization that had split the Ciancaglini family reflected the broader turmoil that was roiling throughout the Philadelphia underworld at the time. On this particular morning, March 2, 1993, what was about to happen to Joey Chang would add substantially to the chaos.

It was still dark when he arrived for work.

The luncheonette was located on Warfield Street, just off the corner of Wharton, in a mixed residential and industrial neighborhood in the Grays Ferry section of South Philadelphia. During the day, the area is heavily trafficked. But before dawn it is desolate.

The surveillance camera picks up the story.

Two people get out of the car: Ciancaglini and Susan Lucibello, a waitress who usually rode to work each morning with her boss.

At 5:56:15 they approach the front entrance on foot. They are little more than shadows on the FBI monitoring screen that is picking up the feed from the pole camera. But the routine is similar to what happens each morning. Ciancaglini reaches down and unlocks the security grate that covers the front of the squat, narrow cinderblock building. Then he opens the front door and flicks on a light that will provide an eerie backdrop for what is to follow. He and Lucibello walk into the restaurant and begin the business of preparing for the breakfast customers—construction workers and warehouse attendants, gas station operators and office workers—who will soon be arriving for coffee, toast, muffins, and, occasionally, a platter of eggs, scrambled or over easy. Most of the trade is take-out, but there are those who grab something to eat at the counter before heading off to work.

At 5:58:18 a station wagon drives past the restaurant, traveling from right to left on the television screen. In the dim predawn light, it is impossible to determine the make or the color of the vehicle. Inside it are the men who are coming that morning to kill Joe Ciancaglini. A sedan, with a lone driver, follows the station wagon. Both disappear to the left, off the screen.

At 5:58:40, three or four shadowy figures—it's difficult to tell—come running from the direction of the station wagon and burst through the front door. The FBI bug that has been planted in the restaurant picks up the next five seconds.

Now there is audio to accompany the video. Susan Lucibello screams; there is the sound of rapid footfalls; then at least two of the shadowy figures disappear into the back storage room where Joey Chang is getting ready for what he had assumed would be just another workday. There is the staccato sound of gunfire, six or seven shots. More screams from Lucibello. The rapid shuffling of feet as the shadowy figures head out the door. A man's voice yelling "Move, move!" as the gunmen exit and disappear offscreen to the left, toward the station wagon.

The tape is stunning. It may be the only time in the FBI's long and storied history of battling organized crime that it was able to record a mob hit in progress. That it happened in Philadelphia makes it even better. Because anyone interested in understanding what has happened to the American Mafia over the past twenty years, anyone who tries to discern how and why this once highly secretive and criminally efficient organization has come undone, must look hard at the City of Brotherly Love.

The demise of the American mob starts here.

The attempted assassination of Joseph Ciancaglini Jr.— miraculously, he survived the hit—comes in the middle of the story, but it is the perfect jumping-off point.

Jack Newfield, the highly regarded New York writer and investigative reporter, had a piece in *Parade* magazine not long ago that asked, in bold headlines, "Who Whacked the Mob?" With all due respect to the federal prosecutors and FBI agents who have developed tremendous cases against La Cosa Nostra, the real answer is simple: the death of the American Mafia is the result of self-inflicted wounds. Call it suicide by arrogance, incompetence, greed, and stupidity.

And don't underestimate the impact of assimilation.

There are many Italian-American groups in the country today who get their noses out of joint because of the popularity of HBO's contemporary mob series *The Sopranos*.

The highly acclaimed show, they contend, "marginalizes" Italian-Americans and reinforces the stereotype that they're all gangsters.

The fact of the matter is, the best and the brightest in the Italian-American community are doctors and lawyers, professors and artists, actors and athletes. From Giuliani to Giambi, from Scalia to Scorcese, Italian-Americans are found at the top of almost any field of endeavor.

The mob is another matter. A couple of generations ago, the guys who ran the rackets had smarts. Take Carlo Gambino in New York or Angelo Bruno in Philadelphia: given the opportunity, they could have run a Fortune 500 company. Not so with the guys who came after them. And the guys running the families today? Fuhged-daboudit.

"They're scraping the bottom of the gene pool," a Philadelphia defense lawyer said during a recent racketeering trial in which his client, a hapless hit man, was convicted.

In fact, *The Sopranos* is dead-on accurate. And what it has captured is the disintegration of a once highly efficient organization.

Honor, loyalty, and a sense of family—noble traditions bastardized by the Mafia to justify its existence—disappeared at least an underworld generation ago. In their place are those classic American values that fuel Wall Street and politics—greed, power, and self-aggrandizement. One of the reasons the mob is dying is that it has become too Americanized. Its value system has been corrupted. In the 1980s, guys like John Gotti in New York and Nicky Scarfo in Philadelphia turned the organization on its ear. They confused fear with respect and, like so many other "leaders" in American society, used the dollar as the benchmark for determining success, and the spotlight of the media as validation of their worth.

"I always wanted to be a gangster," Gaetano "Tommy Horsehead" Scafidi told a jury in Philadelphia during Merlino's big mob racketeering case in the summer of 2001.

Scafidi, who admitted to being one of the gunmen in the Joey Chang hit, epitomizes the new mob. He is a fourth-generation South Philadelphia wiseguy. His great-grandfather had been a capo—a mob lieutenant—at the turn of the last century. His grandfather, two uncles, and an older brother were also made members of the Philadelphia crime family. His brother, Salvatore, was sentenced to forty years in prison following a racketeering conviction in 1988. Still, Tommy Horsehead wanted in.

Not the brightest light, Tommy once was asked by a judge who was about to sentence him in an extortion case what year he had graduated from high school.

"Senior year, your honor," Horsehead replied.

And people wonder why the American mob is in the state it's in?

"My family were wiseguys," Scafidi said while testifying for the government about the Ciancaglini shooting. "I just liked the life-style. I liked the respect. . . . I could go into restaurants, I could get special treatment. . . . We were allowed to do whatever we wanted. We could go into clothing stores and tell the people, 'We want to pick out suits,' and we'll tell them that we'll pay them at the end of the month. . . . [They never did.] We were allowed to go shake down drug dealers, bookmakers, loan sharks, anybody who was doing anything illegal. We went . . . and told them they had to do the right thing. And if they didn't . . . especially people in our neighborhood, they knew they would have a problem with us. . . . They would have gotten a beatin'. Or, if they own a club, we could have wrecked . . . sent people in to wreck the club and things like that. . . . They said they respected us, but they feared us because they knew what we were about."

On the morning that Joey Chang got popped, what the Philadelphia mob was about—at least on the surface—was reorganizing. During the Scarfo years, from 1981 to 1989, the crime family was rocked by a series of prosecutions that were highlighted by the turncoat testimony of two members of the organization, Thomas "Tommy Del" DelGiorno and Nicholas "Nicky Crow" Caramandi. This came at the end of a bloody decade of internecine turmoil in which at least thirty mobsters were killed.

Those convicted and sentenced to lengthy prison terms included Scarfo and most of his top associates, including Joey Chang's father and Scafidi's brother. In all, more than twenty mob members were sent to jail. This in a crime family that, at its peak in the 1950s, had boasted about eighty formally initiated members. Before the feds were finished, three other major figures in the Scarfo organization, including Phil Leonetti, Scarfo's nephew and underboss, would also agree to testify.

So much for family ties.

Omerta, the Mafia's once sacrosanct code of silence, was shattered. In its place was an aria sung by a South Philadelphia boys' choir, a chorus of informants who gave investigators an unprecedented view of life inside an organized crime family. What they saw was more soap opera than *Godfather,* an organization beset with petty squabbles and senseless bickering that often led to bloody beatings and gangland shootings.

Leonetti, offering a classic line that could have been the signature phrase for the Scarfo years in Philadelphia, was once asked on the witness stand whether he considered himself ruthless. The handsome young mobster, who hated the nickname "Crazy Phil" that an Atlantic City radio reporter had created for him, had admitted his own involvement in ten mob hits.

"I know what it means to be ruthless," he said. "But I don't remember ever doing anything, as a matter of fact I know for sure, I never did nothing ruthless besides, well, I would kill people. But that's our life. That's what we do."

Four years after Leonetti made that statement, that's still what the Philadelphia mob was doing. But it had become much more personal, and even more treacherous. On a certain level, it was almost inexplicable.

The Joey Chang hit ended a fragile alliance that had been brokered between John Stanfa, the Sicilian-born Mafioso who had taken control of the Philadelphia family, and a group of young wiseguys headed by Skinny Joey Merlino and Mikey Chang Ciancaglini. Joey Merlino was the son of Salvatore "Chucky" Merlino, a jailed Scarfo lieutenant. Mike Ciancaglini, the youngest of the three Ciancaglini brothers, was also the most violent.

Both Scafidi and federal authorities contend that Mike Ciancaglini orchestrated the hit on his brother Joe, sending the crew of gunmen to the Warfield that morning. Scafidi, in fact, said that Michael told him he would kill Scafidi if he didn't go along with the plan.

The reason Michael wanted his brother dead?

A year earlier, almost to the day—March 3, 1992—two shotgun-toting hit men had chased Michael Ciancaglini through the streets of his South Philadelphia neighborhood. The ambush occurred a little after 8 P.M. Mike Ciancaglini was coming back from playing basketball in a nearby park. Fortunately, he spotted the gunmen before they could open fire. He took off running, and had just ducked into the door of his row house when the shooting started. A spray of gunfire ripped through the door and front windows as Mike Ciancaglini sprawled on the floor. Ciancaglini's wife, Monique, was sitting on the couch at the time. Their two young children—a son, three, and a daughter, two—were asleep upstairs.

Michael Ciancaglini later told Scafidi and others that he was certain one of the shooters was his brother Joe.

"He swears his brother tried to kill him," Scafidi told authorities when he first started cooperating.

Now, a year after that botched hit, Mike Ciancaglini was looking for revenge. The hit men who went to the Warfield that morning expected both Stanfa and Joey Chang to be in the restaurant. Stanfa owned the building, along with a food distribution warehouse just down the street. The warehouse, Continental Food Distributors Inc., was Stanfa's headquarters. He was there almost every day. And almost every morning he would stop at the Warfield first to see his new underboss.

"We're gonna go kill that greaseball and we're gonna go kill my brother," Mike Ciancaglini told Scafidi that morning. "If you don't wanna do it, I'm gonna kill you right here, right now."

Scafidi cringed at the thought of one brother killing another. But he says he had no choice.

Afterward, he said, it was difficult to determine what had upset Mike Ciancaglini more, the fact that his brother Joey had survived the hit, or the fact that Stanfa wasn't there when the gunmen burst into the restaurant.

The night after the shooting, Scafidi had it all laid out for him. He met Joey Merlino at a neighborhood restaurant, DeMedice's, on South Eighth Street. Merlino was at the bar, drinking.

"Michael's fucking furious," Merlino said. "Stanfa wasn't there. His brother's not dead. We hadda make up a story."

Scafidi nodded, waiting to hear more.

Merlino and Mike Ciancaglini, trying to cover their asses, had rushed to the hospital that morning to be at Joey Chang's bedside. Stanfa and some of his associates were already there. Everyone speculated on where the hit had come from. Merlino and Mike Ciancaglini threw suspicion toward a young Sicilian mobster named Biaggio Adornetto, who a few months earlier had had a falling-out with both Stanfa and Joey Chang. Adornetto had worked at the Warfield.

Merlino and Mike Ciancaglini told Stanfa they would help find and kill Adornetto.

"Stanfa bought it for a while," Scafidi said. "A couple of months later, he figured it out."

*   *   *

The audio- and videotape of the Joey Chang shooting is one item in a library of tapes and videos that chronicle the demise of the Philadelphia mob, arguably the most dysfunctional crime family in America. Only the Boston family—with the FBI-sanctioned criminal exploits of Whitey Bulger—is in contention for that dubious distinction. What's more, Philadelphia has been without doubt the most recorded mob family in the country. At the time of the Joey Chang shooting, the feds were already in the midst of a major taping operation that had targeted Stanfa and his top associates.

Most damaging were the bugs they had planted in the offices of a Camden, New Jersey, defense attorney who represented several mobsters. John Stanfa went there often to hold mob meetings, and for two years beginning in 1991, the feds got it all. More than two thousand conversations were recorded. Wiseguys from Philadelphia, Newark, New York, and Scranton were picked up discussing hits, misses, and the generation gap that was tearing the mob apart—a generational divide typified by the clash between Stanfa and the faction headed by Merlino and Mike Ciancaglini.

But that is only a part of the massive audio library that now exists. A year earlier, in 1990, mobster George Fresolone, a member of the Newark branch of the Philadelphia mob, capped a thirteen-month undercover operation for the New Jersey State Police by wearing a body wire to his own mob initiation ceremony.

Fresolone recorded more than four hundred conversations that resulted in the indictments of forty-one mob members and associates from six different crime families operating in New Jersey. Those tapes were so good, so incriminating, that after learning what they contained, thirty-nine defendants pleaded guilty rather than go to trial. Of the two who opted not to plead, one, a mob boss, was convicted. The charges were dropped against the other, a low-level mob associate.

Then came the law-office tapes that targeted Stanfa. And even as the Stanfa investigation was ending, the feds were up and running again, in a probe that would focus on soon-to-be mob boss Ralph Natale, Joey Merlino, and the group of young mobsters who flocked to Merlino's side. These were the mobsters for the new millennium, the me-first, Generation-X gangsters who liked making headlines as much as they liked making money. Using John Gotti as their model, Merlino and his crew brought the Philadelphia mob out of the shadows. Media-friendly and quick with a sound bite, they became celebrities in their own right, combining a high-profile lifestyle with an expanding career in crime.

All the while, the tapes were running. And with Ron Previte's help, they would keep running all the way into 1999.

But that's getting ahead of the story.

By the end of April 1993, seven weeks after Joey Chang was gunned down, John Stanfa was picked up on one of the tapes in his lawyer's office plotting the murders of Joey Merlino, Mike Ciancaglini, and a third wiseguy, Gaeton Lucibello, the husband of the waitress who had witnessed the hit. Stanfa believed that those three, and not Biaggio Adornetto, were responsible for the attempt on Joe Ciancaglini.

Stanfa planned to hold a sit-down in a restaurant in Northeast Philadelphia to discuss, among other things, the failed attempt to locate and kill Adornetto. The idea was to lure Merlino, Ciancaglini, and Lucibello to the meeting and then gun them down.

"I'm a *greaseball*?" Stanfa asked bitterly in a conversation with mobster Sergio Battaglia that was secretly recorded on April 29.

"You gotta hit them when they don't expect no problem," Stanfa continued in his heavily accented English. "See, you no gotta give a chance. . . . You no gotta give no fucking chance. . . ."

Then he formed a gun with his fingers and pointed it at the back of Battaglia's head.

"Over here," Stanfa said, "it's the best. Behind the ear."

Earlier in the same conversation, he and Battaglia had talked about how they would dispose of the bodies. Stanfa, a stonemason and cement contractor by trade, talked of quick-drying cement and bodies buried at a construction site. But he wanted to send a message first. He was particularly incensed with Lucibello, an ally he thought had betrayed him by switching sides and working with Merlino and Mike Ciancaglini.

After he killed Lucibello, Stanfa said, he wanted to "get a knife . . . we'll cut out his tongue and send it to the wife. . . . We put it in an envelope. We put a stamp on it."

"That's good," Battaglia said.

"Honest to God," insisted Stanfa.

"That's it," said Battaglia enthusiastically. "I think it's good."

Often lost in the talk of mob hits and murder plots is the reality of what happens in a gangland shooting. Too often jury members hearing testimony at a trial—or the public reading about it in a newspaper—get a sanitized glimpse of the mess. The police report from the crime scene is matter-of-fact in style and content. The details of an autopsy are dry and scientific. And even the boasts and braggadocio of somebody like Stanfa plotting a murder seem somehow divorced from the reality.

Caught on grainy videotape, the Joey Chang hit offered something different, something that cut much closer to the bone.

Not for nothing is murder called "wet work." When the government presented its evidence in the racketeering case that played out in Philadelphia in the spring and summer of 2001, one of the most compelling witnesses was the emergency medical technician who arrived on the scene at the Warfield shortly after 6 A.M.

Her name was Colleen Mitchell. By the time she testified, she was an officer with the Philadelphia Fire Department. But on March 2, 1993, she was a first-year EMT working with an ambulance squad. There were three or four police officers already working the hit when she arrived there that morning, she said. One was interviewing Susan Lucibello in the restaurant. Another told Mitchell that there was a shooting victim in the back room.

Mitchell, trained to respond to the most serious injury first, rushed into the restaurant. What followed, she said, she will never forget.

"The woman at the counter [Lucibello] was upset," Mitchell recalled. "I was briefed by a female police officer who said the person inside [Ciancaglini] was dead. But when I went back there, Joseph Ciancaglini was sitting up. He was in an upright position, with his back toward me. I looked at him and at that point I saw all his injuries."

Ciancaglini, the young, handsome, dark-haired mob leader, was sitting in a spreading pool of his own blood. Sticky and dark red, it was pouring out of his mouth, out of his nose, and out of the bullet wounds that she could see in his head, his cheek, and his shoulder. He had also been shot in the foot and leg.

His left eye was swollen shut. There was a jagged wound around his ear, part of which had been blown off. There was another exit wound in the cheek area—a "ragged" wound, she noted. And an entrance wound in the left temple area. He was mumbling and then coherent, lucid and then out of control.

Mitchell took a towel and tried to stop some of the bleeding. Joey Chang grabbed it and tried to blow his nose. She yanked it away from him, afraid that he might literally blow his brains out.

Looking back on it eight years after the fact, she said the events played out like a "silent movie."

"He mumbled a lot," she said. "But at one point he spoke

clearly. It gave me a chill down my spine. I felt he was trying to tell me something, and that he was dying."

On the witness stand, Mitchell said she could not remember anything specific that Joe Ciancaglini said that day. Even
when defense attorneys showed her the police report of the
statement she gave immediately thereafter—in which she
reported that Joey had told her "Tim did me"—she said she
couldn't remember. Defense attorneys would argue that
"Tim" was a nickname for Biaggio Adornetto, an argument
that seemed to have little basis in fact.

But the legal ramifications of Mitchell's testimony were
secondary to the emotional impact it had in the courtroom.

"I never saw anybody so injured have the incredible strength
that man did," Mitchell said as she described Joey Chang's
fight to survive. "He was fighting us, and when the paramedics
arrived they had to handcuff him so that they could get him on
the stretcher. . . . I was trying to keep him from hurting himself. It was the most blood I have ever had on me."

Joe Ciancaglini Jr. survived the murder attempt, but was
never again a factor in the mob. He has lost his sight in one
eye. His hearing and speech are impaired, and he needs a
cane to walk. His ability to remember what happened has
never been tested; neither the prosecution nor the defense
called him as a witness in the racketeering case.

After the FBI heard Stanfa plotting with Battaglia to avenge
the Joey Chang shooting by murdering Merlino, Mike
Ciancaglini, and Gaeton Lucibello, the feds shifted their attention to the young wiseguys. It is FBI policy to notify anyone who is the target of a murder plot that has been
uncovered during an investigation. Agents visited Merlino
and the others and told them they had a problem. The agents
said they couldn't provide any details, but they did offer all
three the option of cooperating. It was an offer Merlino,
Mike Ciancaglini, and Lucibello refused.

Over the course of the summer of 1993, both sides loaded

up. Armed hit men cruised South Philadelphia looking for a chance to open fire. Merlino and Mike Ciancaglini set up a clubhouse in a brick storefront at Sixth and Catharine Streets that used to be the headquarters of the Philadelphia branch of Greenpeace. Once a locale where dedicated do-gooders discussed saving the rain forest, it was now the spot where Merlino, Mike Ciancaglini, and a dozen other young wiseguys would meet each day to play cards, watch tele-vision, and plan their takeover. At the time, they were in constant touch with Ralph Natale, an associate of legendary Philadelphia mob boss Angelo Bruno.

Bruno had been killed in 1980, a murder that many be-lieve set the Philadelphia mob in a free fall from which it was never able to recover. Natale, a former official with the mob-tainted Bartenders Union, was doing time for drug trafficking and arson. He had been in jail since 1979, but hoped to be paroled in 1994. In his late fifties, he had met Merlino in jail two years earlier. Now he was encouraging the young wiseguys. He promised them support from the New York families with whom he had made connections during his decade and a half behind bars. He boasted about how he, Merlino, and Mike Ciancaglini would put the Philadelphia family back in order. He argued that Stanfa—a "greaser" from Sicily—had no right to head an American mob family.

The feds knew about a lot of this because they had been listening to recordings of telephone calls Natale made from prison. Stanfa also learned about the plots. At one point he said of Natale, "As soon as he comes out, he's gotta go. I don't even wanna give him time to breathe the air."

But first Stanfa wanted Merlino and Mike Ciancaglini.

He got his chance on August 5, 1993. The two young mobsters were walking toward their clubhouse when two Stanfa hit men, John Veasey and Phil Colletti, spotted them. They drove by and opened fire. It was a little before 1 P.M. on a hot and humid summer afternoon.

Mike Ciancaglini died on the sidewalk that day, shot once through the heart. Joey Merlino, whose arrogance and bravado were matched only by his luck, took a bullet in the ass. He was walking with a cane within days of the hit, and by the end of the month was back on his feet.

A day after the shooting, the FBI heard Merlino explain to Natale what had happened.

"We both went down together," Merlino said. "I knew I wasn't hurt that bad. It looked like he got shot in the arm."

But as he and his boyhood friend lay on the sidewalk, Merlino said, he could tell from Mike Ciancaglini's eyes that he was in trouble.

"I'm dyin'," Ciancaglini told Merlino. As he said it, blood poured out of his mouth.

"It was just bad luck," Merlino told Natale. Mike had seen the gunmen, and threw up his arm to protect his face; the bullet that killed him entered his body under his armpit and pierced his heart.

"Motherfuckers," Natale said. "Dogs. Cowards."

"I told his wife," Merlino said. "I said, 'I wish I would've died; he would've lived. I got no kids.'"

Then Merlino reminded Natale of how they used to joke about life and death in the underworld when they were cellmates.

"You used to laugh when I told ya," Merlino said, "'They can't kill me. I got the devil in me.'"

The shootings were the seminal event in the Merlino-Stanfa mob confrontation. Now it was open warfare. Natale, cheerleading safely from the sidelines, encouraged Merlino to shoot back, to "do some work" to avenge the death of Mike Ciancaglini.

"Everybody feels that a fine young man just was killed . . . lost his life for nothing," Natale said in one conversation. In another, he told Merlino, "If I was on the street, forget about it. I wouldn't care if the law knows. . . . I'm gonna be ready to do what I gotta do."

In one call from prison, however, Natale ended up on the phone with Merlino's older sister, Maria, and the conversation took a decidedly more realistic tone. Maria Merlino, a one-time Mafia princess who had seen a former fiancé gunned down gangland-style several years earlier, had had enough of the wanton and senseless violence.

"They're all fighting and over what?" she asked. Then, without a pause, she answered her own question.

"Jail time or coffins," Maria Merlino said.

2

The trouble had started years earlier.

The root of it all went back to November 2, 1976, the day voters in New Jersey approved casino gambling for Atlantic City.

The fading resort had always been part of the Philadelphia mob's territory. Before the coming of the casinos, though, nobody cared. Once a wide-open city full of fancy nightclubs and backroom casinos, Atlantic City had hit the skids by the 1960s. Fancier vacation destinations, cheap airline flights to the islands, and the allure of the sparkling casinos in the Nevada desert all contributed to its demise.

That changed when 1.3 million New Jersey voters said yes to legalized gambling on the Atlantic City Boardwalk in November 1976. The voters—a 57 percent majority—accepted the arguments casino advocates had made during the hotly contested referendum campaign, a campaign funded in part by mob-tainted labor unions and supported by mob-linked politicos.

The pitch was simple: The slot machines, roulette wheels, and blackjack and craps tables weren't an end in themselves, said proponents of the plan, but a means to an end. Legalized gaming, said the pro-casino forces—they loved to use the word *gaming* rather than *gambling,* confident that dropping those two letters would eradicate the stigma—held the promise of a "unique form of urban renewal" for Atlantic City. God knows all the other forms had failed. In the mid-

1970s, the city was one of the most depressed urban centers in a state whose cities were cesspools of crime, corruption, and decay. Atlantic City was Newark with a boardwalk, Camden with a beach.

And so the experiment began.

Resorts International—with its questionable ties to Meyer Lansky–linked wiseguys in the Bahamas—opened the first casino in May 1978. Then came Caesars World, with past connections in the Poconos to another Lansky crony. Next was Bally's, the slot machine giant whose New Jersey roots went back to Newark wiseguy Gerry Catena.

Of course, lawyers for those casino companies argued at licensing hearings that their corporate clients had long since severed any ties to organized crime. The proposition that legalized gambling would bring the mob to Atlantic City was ludicrous, they said.

And they were right. Casinos wouldn't bring the mob to Atlantic City: it was already there. For years, Nicodemo "Little Nicky" Scarfo had been the caretaker for Angelo Bruno's interests at the shore. Like the city in those pre-casino days, though, Scarfo was down and out.

"He had a little sports book, he was sharking some money, and he had a piece of an adult bookstore," mob informant Nick Caramandi said of Scarfo in the days before legalized gambling. "But he was dead-out. There was nothin' goin' on down there. He got sent down there as punishment 'cause he stabbed a guy in a diner in South Philly. It was like he was banished."

The stabbing had occurred in the early 1960s. Scarfo, a mob soldier and amateur boxer, got into an argument with a longshoreman over who was entitled to a booth in the Oregon Diner, the same diner where Previte and John Ciancaglini had their sit-down thirty years later. Words were exchanged. There was some pushing and shoving. Scarfo grabbed a steak knife and stabbed the longshoreman in the chest.

Scarfo did about two years on a manslaughter charge. When he got out, several members of the organization wanted him killed. Scarfo was a hothead, they said; he would only bring trouble. At the time, no one realized how accurate they were. Bruno could have saved the organization a lot of aggravation if he'd gone along with the idea of whacking Little Nicky. But Scarfo had two uncles who were capos in the Bruno organization. They got him a pass. He was sent to Atlantic City instead of the morgue.

He languished there for more than a decade—until the casinos came back.

Legalized gambling turned Atlantic City into a boomtown. Real estate values went through the roof. Developers flooded the city with cash. There were jobs and money and tourists and action, all kinds of action.

For a time, Scarfo was one of the biggest winners in town. Power, money, status, and prestige—all the things he had wanted in his struggling days—were right there in front of him. In ten years, he blew it all.

"Atlantic City, that was the beginning . . . of all those troubles," Natale would say twenty years later.

Angelo Bruno was killed in March 1980. There are two theories about why he got popped, but little debate about who was behind the hit. The murder was set in motion—and might even have been carried out—by Antonio "Tony Bananas" Caponigro, Bruno's consigliere.

Caponigro and a faction of the crime family that backed the hit were fed up with Bruno's old-fashioned ways. They thought he wasn't taking advantage of all the opportunities casino gambling presented. They also thought he was allowing New York mob members too much play. New York wanted Atlantic City to be like Vegas, "open" to all mob families. Caponigro and those around him argued that Atlantic City had always been part of Philadelphia's domain. As far as he was concerned, anybody else who was trying to do business in Atlantic City had to do two things: get the okay of the Bruno organization, and give the Bruno family a piece of the action.

Caponigro also found Bruno's policy on narcotics hypocritical and self-serving. Officially, no one in the family was allowed to traffic in drugs. Caponigro and several others had deals on the side, but they had to operate in secret. If they were caught, they could have been killed.

At the same time, however, Bruno was allowing Rosario and Giuseppe Gambino, distant Sicilian cousins of his good friend Carlo Gambino, to operate freely in South Jersey. They had opened a nightclub in Cherry Hill called Valentino's in the early 1970s and had a group of friends and associates spread throughout the Philadelphia suburbs operating pizza parlors. In fact, they were part of the infamous Sicilian "pizza connection," an international heroin ring that made millions dealing smack out of pizza shops throughout the country.

Caponigro knew that the Gambinos were paying tribute to Bruno in order to do business in the Philadelphia area. To Caponigro, that meant that Bruno was benefiting from the drug trafficking. Yet Bruno was forbidding his lieutenants and soldiers to partake in the financial windfall that drugs could bring.

On the night of March 21, 1980, as Bruno sat in the passenger seat of a car in front of his South Philadelphia row house smoking a cigarette, a man in a raincoat walked up from behind, leveled a shotgun at the back of Bruno's head, and pulled the trigger. Many people believe Caponigro was that man.

"He was the kind of guy who would do it himself," said the late George Fresolone, a Newark mobster who knew Caponigro well. "Most of us thought he was the shooter."

John Stanfa, Bruno's driver that night, was slightly wounded in the shooting. To this day there is underworld and law enforcement debate over whether Stanfa was in on the murder plot. He disappeared shortly after the shooting and a year later was found operating a pizzeria in Maryland. He was brought back to Philadelphia, where he was convicted of perjury for lying to a grand jury investigating the Bruno murder. He did eight years, safely tucked away in a federal pen while the Philadelphia mob began to self-destruct.

Tony Caponigro didn't have much time to capitalize on Bruno's death. He and three of his top associates turned up dead within months of the hit, brutally beaten and shot, their bodies dumped where they could be found. Caponigro's body was left in the trunk of a car in the South Bronx. Several twenty-dollar bills had been stuck in his ass, a signal that he had gotten too greedy. Murdering a mob boss at the time was still very much frowned upon.

Phil "Chicken Man" Testa, Bruno's underboss, took over the organization. Testa, whose family had once been in the poultry business, tapped his long-time friend Nicky Scarfo as his consigliere. Within a year Testa was killed, blown up by a nail bomb planted under the porch of his home. His death closed the first act of a power struggle in which Atlantic City was the prize. This wasn't a dispute over South Philadelphia bookmaking and loan-sharking operations. These guys were killing one another over the pot of gold they all saw sitting on the Boardwalk.

Everyone knew the stakes had gotten higher. All of a sudden the Philadelphia mob, the low-key, backroom organization Angelo Bruno had nurtured for years, was becoming a high-profile presence. Bruce Springsteen, The Boss himself, wrote in 1982 about the Testa hit in his song "Atlantic City," lamenting how they blew up "the chicken man."

Scarfo, with the backing of the Genovese family in New York, took over the top spot. He avenged Testa's death by ordering the murders of two of the prime suspects in that hit, and then rebuilt the Philadelphia crime family in his own image.

Angelo Bruno had been a Mafia diplomat, very much an

old-world don. He believed in making money, not headlines. He didn't have a flamboyant bone in his body. He bought his clothes—conservative business suits, white shirts, striped ties—off the rack. He drove a Buick.

Nicky Scarfo brought the Philadelphia mob out of the shadows. He was John Gotti before there was a Gotti. Fancy clothes. Custom-made shirts. Italian leather shoes. A Mercedes-Benz. Ringside seats at the prizefights staged by the casinos. A forty-one-foot yacht he called *Usual Suspects* was docked behind his lavish vacation home in Fort Lauderdale, a home he called Casablanca South. (Scarfo loved Bogart films; *Casablanca* was his all-time favorite.)

"Look at me, I'm a gangster," Scarfo said with everything he did.

Atlantic City gave him the perfect stage.

By the early 1980s, the city was up and running. There were six casino-hotels, then eight, then ten, finally twelve. Scarfo got his hooks into the Bartenders Union, which represented most of the casino workers. Before the feds moved in to oust the mob-tainted union leadership, Little Nicky was taking thirty to forty grand a month out of the union's coffers.

The money came in cash, mostly ten- and twenty-dollar bills, delivered by a union official to Scarfo's home on Georgia Avenue in Atlantic City. Then Scarfo would send his son, Nicky Jr., and Tommy Scafidi to one of the casinos to "wash" the money.

"Scarfo liked hundreds, he didn't want all these tens and twenties," Scafidi recalled later when describing their monthly routine. He and Nicky Jr. would take the small bills down to one of the casinos on the Boardwalk. Caesars Boardwalk Regency and Bally's Park Place were within three blocks of Scarfo's residence. They would buy chips and pretend to gamble. Play a few hands. Walk the casino floor. Then they'd cash out and bring the money—nicely converted to neat, crisp stacks of hundred-dollar bills—to Little Nicky.

It was just one of the many benefits casino gambling brought to the mob. The gaming halls were a place to launder money, a place to hang out, and the perfect venue for almost any mob gambit ever conceived.

Prostitution, illegal gambling—poker and sports betting were still against the law—and various con games all became a part of the new casino underworld. Scarfo was sitting on top, but there was so much going on—so much money being made—that even he couldn't keep track of it all.

Atlantic City was a hustler's wet dream. Hundreds of wiseguys and wannabes flocked there to experience the rush. It was a chance for anybody on the make to take a shot. Most failed and went broke. They got caught up in the pitch, the lure, the fantasy. They couldn't stay away from the tables. For the four, six, or eight hours they were gambling, they escaped the humdrum nine-to-five world of suckers and losers. They could be whoever they wanted—Frank or Dino, Pacino or De Niro—as long as they were willing to pay the price. They gambled away their paychecks. They mortgaged their houses. They busted out chasing the illusion, pretending to be what they weren't.

Others, however, saw the gambling experiment for what it was—an opportunity.

Put Ron Previte in that group.

Pick any mob scam—drugs, prostitution, bookmaking, extortion—and Big Ron Previte has had a piece of it. Need a horse race fixed? He'll tell you how. Want to know how to rip off a casino? He can show you twelve different ways.

His life was crime: organized and disorganized, petty or high-profile.

"Every day was a different felony," says the man who eventually brought down the Philadelphia mob of Ralph Natale and Joey Merlino.

First as a Philadelphia cop, then as a freelance underworld

entrepreneur, and finally as a made member and capo in the organization, Previte left his fingerprints on millions of dollars.

And make no mistake, no matter what role Previte was playing, it was always about the money.

As a cop he was corrupt. As a mobster he was cunning and at times brutal. But it was in his final underworld role, as a confidential informant and government witness, that the six-foot, three-hundred-pound gangster was deadly.

His time in Atlantic City set the stage for all that. It's when Previte first came in regular contact with wiseguys, when he first met them on their terms. As a cop, he'd dealt with mob guys, taken money from them. But in Atlantic City it was different. Now he was a player. He wasn't looking for a bribe or a kickback, he was looking for a piece of the action. He was setting his own deals in motion, bobbing and weaving in an underworld full of schemers.

Atlantic City is also where Previte first realized that he was as smart as, maybe smarter than, most of the mobsters he had to deal with. Previte was a crook long before he was a gangster. More important, he was always a mercenary.

"I've always looked out for myself," he said during one of several candid conversations in which he explained his philosophy of life. "Nobody else is going to. Look, you do serious things, there are serious consequences. I always knew that. But I also knew if you wanted to make money, you had to take risks. Fortune favors the bold.

"If I fell out of this chair right now with a heart attack, I'd have no regrets," he said in the summer of 2002 as he sat in a South Jersey restaurant working on a plate of clams and spaghetti. "I lived my life the way I wanted to live it.

"Didn't make all the money I wanted to make," he added with a laugh, "but I made a lot."

Previte grew up in West Philadelphia, where his father, Jesse, a Sicilian immigrant and decorated World War II veteran, ran a gas station. His mother's family came from the area around Naples. He smiles and tells the story of how her family was horrified when she announced she was going to marry a Sicilian.

"My father is a hard-headed Siggie," Previte says of his dad, a one-time amateur boxer with hands "like shovels" who fought under the name Jesse James. "My mother is the sweetest woman you'll ever meet. . . . Sometimes I don't know how she put up with him all these years."

An altar boy at Our Lady of Lourdes grammar school, Previte spent his first year of high school in a local seminary. Long before the pedophile scandals that rocked the Catholic Church, he knew there was something wrong among the priests. The seminary "was a haven for homosexuals," he said. He left in less than a year.

"My grandmother wanted me to be a priest," he says, shaking his head and rolling his eyes at the memory. "I did it for her. But I knew it wasn't gonna happen."

Previte's family moved back and forth between Philadelphia and Hammonton, New Jersey, while he was growing up. Located in Atlantic County, about thirty miles west of Atlantic City, Hammonton happily describes itself as the "blueberry capital of the world." It is a small, tight-knit community, the kind of place Norman Rockwell might have painted if he spoke the Sicilian dialect and appreci-ated the delicacy inherent in a perfectly prepared plate of broccoli rabe.

Many of the twelve thousand residents of Hammonton traced their roots to a group of Sicilian immigrants from the village of Gesso, who came to Hammonton just after the turn of the century to farm and make a better life for them-selves. Their struggles were chronicled in a report called "The Italian on the Land," written in 1907 by Emily Fogg Mead, who came to study the new immigrants for the U.S. Department of Labor. Emily Mead's daughter, Margaret, followed in her mother's footsteps and became a renowned anthropologist and author.

Previte's father's family was part of that group. "Hillbilly Sicilians," he called them. They were farmers and later businessmen, fiercely independent; men who, no matter how much money they had stashed away, still drove pickup trucks and made their own wine. Hell-bent on survival and self-sufficiency—almost to the point of ruthlessness—they were epitomized in a not especially flattering Italian proverb: "Sicilians love their money more than they love their children. And they love their children very, very much."

Previte attended Hammonton High School, where he played football, got decent grades, and developed the character traits that would serve him well as an adult.

"He didn't trust too many people," said Joe Bartuccio, a longtime friend and high school classmate. "He hung with a small group of guys. He was generous and above average when it came to intelligence. But he could be cunning and diabolical. . . . Even in high school, he was always trying to stay one step ahead of everybody else."

After high school, Previte enlisted in the Air Force; when he came out, he joined the Philadelphia Police Department.

He left his mark on both organizations.

In the Air Force, he and another airman who worked at a supply depot made thousands of dollars selling government issue on the black market. Leather flight jackets and aviator sunglasses were particularly lucrative and easy to move, he says. He got jammed up, however, and nearly bounced out of the service after he and his partner were caught trying to break into the credit union safe at the base where they were stationed in Mississippi.

"My buddy said he was real good with explosives," Previte says, recalling the incident. "We ended up burning down a couple of buildings, but not the door on the safe."

Nearly a month in the brig, and some intense interrogation, followed.

"They used to strip you naked and throw cold water on you while you were handcuffed to a post," Previte said. "A little while after that, they'd throw hot water on you. They can't do that now, but they did back then."

Previte, who had already done a tour overseas, figured he was headed for a dishonorable discharge. But neither he nor his partner admitted to the attempted heist.

"They knew it, but they could never prove it," said Previte. It was a position he would find himself in again and again as his criminal career expanded.

After spending some time in the brig, Previte had to undergo a series of tests administered by an Air Force psychiatrist.

"After I take the tests, the psychiatrist calls me in for an interview," Previte said. "I'm thinking it's gonna be some kind of counseling session. I'll never forget it. He looks at all these test results and he says to me, 'Son, there's nothing I can do for you. You have a criminal mind.'"

Previte ended his Air Force career with a twelve-month assignment driving a transport bus on a base in Goose Bay, Labrador.

"One of the coldest and most desolate fuckin' places in the world," he says. "We couldn't do anything because there was nothing there . . . which is why they sent me, I guess."

The Philadelphia Police Department was his next logical stop.

"I learned more about being a crook while I was a cop than at any other time in my life," Previte says of the twelve years he spent on the force.

Released from the Air Force with an honorable discharge despite his problems—"They were just happy to get rid of me"—Previte took the test for the police academy to spite his father, who had moved back to Philadelphia and who wanted his son to join him in running a gas station there.

"I wanted no part of that," said Previte.

As a Vietnam era vet, Previte ranked high on the list of candidates for the police academy. But he had a problem. His eyesight had been deteriorating steadily since he joined the Air Force. By the time he took the police exam, he could barely see out of his left eye.

"I go to this eye doctor's office to be examined and I fail the eye test," Previte said. "He says there's no way I can get on the police force. I'm practically blind in one eye. Then he asks me if I'm Jesse Previte's son? I say, 'Yeah.' So he tells me to come back that afternoon.

"I said, 'Doc, how's my eyesight gonna get any better by this afternoon?' He just tells me to come back."

Previte was sitting in the family kitchen a few hours later when his father called. His mother answered.

"He told her to give me five to take with me when I went back to the eye doctor's," Previte said.

Five: that was five hundred dollars, cash. With that, Previte passed the eye exam and qualified for the police academy.

Over the next twelve years, working primarily out of the Twelfth Police District at Sixty-fifth Street and Woodland Avenue in West Philadelphia, he would recoup his "entrance fee" many times over, taking money wherever and whenever it was offered. He shook down drug dealers and pimps, rousted hookers, willingly took envelopes from mobsters to ignore their gambling and bookmaking operations, and stole whatever he could get his hands on.

"One of the first jobs I'm on, we're gonna raid this craps game in a black neighborhood," Previte said. "A couple of old-timers are breaking me in. I'm supposed to wait around the back of the building. The game's on the second floor. They're gonna bust in the front door and chase everybody out the back. I round them up and pretend to arrest them. I hold them for a while, then let them go. But we get all the cash."

Nobody bothered to check around back, however.

The outside stairway from the second floor had collapsed. The gamblers who came running out fell two stories. One guy broke a leg. Another injured his back. Previte and his partners were forced to call in ambulances.

He still remembers the reaction of one of the old-time cops that night. "I guess we gotta put some of the money back," he said.

"Every day there was something," Previte recalls.

"I used to pick up the envelopes from Harry Riccobene," a legendary, old-time Philadelphia mobster. "He was paying off everybody in the precinct so that his gambling and bookmaking operation could operate without any hassles."

Previte, ever the entrepreneur, also had his own little book, taking numbers and sports bets on a pay phone in the hallway of the station house. Occasionally, if the action got too heavy, he would edge off to a bigger book run by the mob.

But there were also days—most days, in fact—when Previte performed legitimate police work.

"I was a good cop, I was good at my job. I was also a crook. But I never shied away from police work and I never took advantage of innocent people. I enforced the law. I ran into burning buildings. I took down drug dealers. I broke up bar fights. I never worried about myself. It's like when you play football. It's the guy who's careful and cautious who gets hurt. You have to be fearless. That's the way I approached the job.

"You know, there are people now who want to say I was nothing but a bum when I was a police officer. That's not true. I was a crook, but I wasn't no bum. There's a difference. Guys used to work that twelve-to-eight shift, they'd be sleeping somewhere by two A.M. I worked. I was always out looking for some action. When it mattered, I enforced the law."

He's got the commendations to back up his claims. In a performance report from May 1970, for example, he was rated superior or outstanding in all ten performance areas, from quality of work and work habits to initiative and dependability.

"You have shown that you can be depended on to do your

work with little supervision," the report read in part. The lieutenant filling out the rating did caution Previte to "keep after your personal appearance"—a reference, Previte says, to his penchant at the time for wearing his hair long.

A year later, he was again rated superior and outstanding. The report said Previte had "promotion potential." It also noted that he had been injured on the job and missed several months of work.

There was, of course, a story behind the injury. "Me and my partner had the airport as part of our territory, and every night we had to drive through this construction area out there," he said. "There were these construction trailers with these big air-conditioning units in their windows. Every night my partner goes, 'I'd like to get one of those air conditioners.' Finally I say, 'All right, I'll get ya one.' "

They were patrolling in a police wagon at the time, one of those large, jeeplike minivans. Shortly after the start of the midnight shift, they pulled the vehicle up under the window where the air conditioner was located. Previte climbed on the roof of the patrol wagon and began to work the air-conditioning unit loose, removing the screws and bars that held it in place. As he was doing this, his partner accidentally shifted into gear and the wagon began to rock.

Previte lost his balance and crashed to the ground.

"I wrecked my fuckin' knee," he said. "The pain was excruciating."

But they couldn't go to the hospital without an explanation for what had happened.

"We hadda wait for a call to come in. I'm sitting in the wagon for two, three hours in pain waiting for something to happen."

Finally they get a radio call to respond to a fight at a bar.

"We rush over there," Previte said. "I limp in and grab a guy who's fighting. When he sees we're the police, he doesn't put up much resistance, but I keep wrestling with him. He's looking at me funny, but I won't let go. Finally a sergeant arrives and I crumble to the ground, hollering about my knee.

"They take me over to the hospital. I had torn all the ligaments. I went out on a job-related disability."

The air-conditioner scam was a lark, but Previte readily admits that the crimes he committed while in uniform weren't always as benign. Almost all of them, he said, involved money. Some involved a large amount of money.

"Guys would be fighting to work with me," he said proudly. "Especially if they had a bill that was due. I could always smell the note. I knew how to find money."

One of his biggest paydays came in the early 1970s, when he and his partner, along with several associates, ripped off "some big-money hippies living in Bryn Mawr," a fashionable suburb on Philadelphia's Main Line.

"My partner had a cousin who had lost an arm in Vietnam, and he was making money driving loads of marijuana up from down south for these rich hippies. We figured we could rip them off. I had this guy Bruce, a black drug dealer, who could move the stuff for us. I used to drive with Bruce when he was making his deliveries up in Germantown. He'd be dropping off a pound at a time. He liked having me around 'cause I had the badge. If anybody stopped him, I'd flash the badge and tell them it was police work. He was paying me three hundred a week to help him with the deliveries. I did this when I was off duty. We never had a problem."

Through the driver, Previte and his associates found out where the hippies were storing their stash. One night they hit the place, a garage out in the suburbs.

"We busted in there, pistol-whipped these two hippies, and took the pot. Musta been two hundred and fifty pounds. Brought it to Bruce. He started selling it."

A month later they learned of an even bigger shipment that the one-armed driver was bringing up from the south. This time they arranged to hijack his truck.

"We had it prearranged. See, the driver wasn't a suspect in the first heist, so he was still working for them. We tell him we're going to be waiting for him at this diner in Lancaster County. He and one of the hippies are driving the load up.

"While they're in the diner eating, we're taking the pot out of his truck and putting it in this van that we had. There was about five hundred pounds. High-grade stuff. We take the van and stash it."

But to make the robbery look legitimate, Previte and one of his associates stayed behind to confront the one-armed driver and the hippie.

"We pull out guns and force them into this car. We're going to drive them out into the country and leave them and then head back to the city. But I make a mistake. I tell the hippie I'm going shoot him. He panics and grabs the steering wheel. The car rolls over into a ditch. This is out in Pennsylvania Dutch country. All farms."

Previte's gun fell on the floor as the car turned over. The hippie grabbed it as they all scrambled out of the vehicle.

"He's got it right at my head," Previte says, holding his fingers cocked like a gun to his temple. "He pulls the trigger and I hear a click. The gun jammed or something. Then I hear this boom. My partner's got his gun out and he's firing at the hippie. Hits him in the shoulder. We leave him there. We musta walked for four hours before we get to this phone booth. I call this guy to come get us."

The next night, while on duty, Previte and his partner took their police wagon to the spot where the van was hidden. They off-loaded the marijuana into the police vehicle and drove it to Philadelphia's South Street, where Bruce had a business.

"We had to use those wooden beer rollers to get the bushels of marijuana out of the van and into the basement of his place. Then he started selling it. It was good stuff. I think we got like six hundred dollars a pound. That was our end. You figure we brought him seven hundred and fifty pounds, that's almost a half-million dollars. Big money."

Previte said he laughed a few days later when he read a newspaper account of an apparent abduction and shooting in Lancaster County.

"This hippie tells the police out there that two guys grabbed him, forced him into a car, and then shot him and abandoned him. What else could he say? He couldn't tell them we robbed his marijuana. The only bad part was, my partner's cousin couldn't drive for them no more. After the second time, they knew he was involved."

On another occasion, Previte said, he and his partner were chasing a stolen car, a 1962 Chevy. They were on Cobbs Creek Parkway in Southwest Philadelphia when the car thief abandoned the vehicle and took off on foot.

"I had to drive the car to the pound," Previte said. "It was an older Chevy, but I loved driving those kinds of cars. As I'm driving, the trunk pops open. I stop and look inside. There are stacks of twenty-dollar bills. Musta been a hundred thousand dollars. I call my partner. He drives over. He looks at the money and then he notices, 'They all got the same serial number,' he says. It was counterfeit."

But this was before businesses were equipped with the blue lights and yellow pens they now use to test the authenticity of currency. For the next three months, Previte and his partner would drive out to Delaware County before the start of their shift. They then worked their way back to the city, stopping several times along the way to buy coffee at a McDonald's or Burger King. Each time, they made their purchase with a twenty-dollar bill.

"We did this for weeks and weeks," Previte said. "I musta had a couple hundred of those white McDonald's coffee stirrers. But we passed all the money."

Previte says he took pride in most of his scams. But there's one that still bothers him. He and his partner were working the area out around Philadelphia International Airport on a Sunday afternoon. There were a lot of dirt roads and secluded spots that they would routinely patrol. This afternoon they came across a station wagon parked off to the side. The windows were steamed up.

"Right away, I figure somebody's in there having sex," Previte recalled.

He was right, but it wasn't the kind of sex he was expecting.

"I look in the window," Previte said. What he saw was a middle-aged man having sex with a young boy.

"I grabbed the guy, pulled him out, cracked him a couple of times. Threw him up against the police wagon. He's crying and asking us to let him go. 'I got a problem,' he says. I tell him, 'You had a problem. Now you've got a catastrophe.'"

Previte talked to the boy, a skinny twelve-year-old, who explained that his uncle took him out to dinner every Sunday. Afterward, he said, they would park and "play."

"I went back and slapped the guy a couple more times. I tell him he's going to jail. He's begging now. He offers us 'twenty-five.' I slap him again. 'Twenty-five hundred?' I say. 'You're going to jail.'

"'No,' he says. 'Twenty-five thousand.'"

And so the negotiations began.

The molester was an executive at a local production plant. An arrest would end his career. His final offer was forty thousand. Said he had the cash in his office; Previte could pick it up the next morning.

"I talk to my partner. He says we're taking a big chance. He figures if we let this guy go he'll call Internal Affairs and say we're shaking him down, and they'll be waiting for me when I get to his office."

Previte, however, figured that was unlikely. How would the executive explain how he'd come in contact with the two officers in the first place? What would the nephew say if he were questioned?

"I told my partner it was big money and it was worth the risk."

Previte cut the deal with the executive, but first he took down the names, addresses, and phone numbers of the executive and his twelve-year-old nephew. "I told this guy I'd be there in the morning. I also told him if he gave us up, he'd have a bigger problem than we would. I told him I would haunt him for the rest of his life."

The next morning, Previte and his partner, off duty and driving an unmarked car, pulled into the parking lot of the executive's company. Previte took an elevator to the office, as he was directed, and was told by a secretary that he was expected. He walked in and was handed a brown paper bag.

"I took the money," he said. "At that point, if Internal Affairs was there, that's when they would have arrested me. I told the guy he was a scumbag piece of shit and I left."

In the car driving away, Previte looked in the bag. The cash was all there.

"We drive a little way and I tell my partner to pull over at this phone booth," Previte said.

Previte got out of the car, placed a quarter in the slot, and dialed a local number. When a woman answered, he told her, "You don't know me, but there's something I want to tell you about your brother."

When Previte finished, he heard the woman scream. Then he hung up the phone.

Two days later he walked into Center City Cadillac at Broad and Ridge Streets in Philadelphia, plunked down seventy-five hundred dollars in cash, and bought a leftover 1973 Cadillac Coupe DeVille.

"It was beautiful. It was a pink color. They called it Mountain Laurel. One of those special Cadillac colors. It was loaded. I had it detailed. Had the Cadillac floor mats, the works. I told the guy Tony, who ran the place, I wanted to drive it off the lot. He said it might take a couple of days. I said I wanted it now. I gave him the money and said I was gonna get some breakfast—have it ready when I get back. I remember they had this wooden ramp. I drove that car down the ramp that day.

"I loved that car. But I still feel bad about how I got the money. That's why I made the phone call.

"It went on like that for twelve years," Previte recalls. "Every day I was committing a crime. Some days more than one."

Part of the culture of corruption that permeated several layers of the Philadelphia police force in the 1970s—bribery and brutality were commonplace—Previte was eventually driven off the job in a departmentwide anticorruption campaign. At the time, he was working in the auto pound, the Police Department's version of Goose Bay; like the Air Force before them, his supervisors had decided it was better to hide Previte off where he could do the least damage. But soon Internal Affairs was forced to open an investigation into the theft of valuables—watches, jewelry, tape cassettes, literally anything that wasn't nailed down—from impounded automobiles that were being stored in police custody. Most of the thefts occurred when Previte was on duty; before long, he "retired" on partial disability.

With his police background, Previte was able to land a job in the security department of the Tropicana Casino Hotel that was about to open in Atlantic City. This was early in 1981.

The Trop was the ninth casino to come online. At the time, it was the most expensive ever built on the Boardwalk. But when Previte went to work there, it was in the late stages of a troubled development. It was behind schedule and tremendously overbudget.

Originally, Ramada Inns Inc. had planned to refurbish the old Ambassador Hotel on the Boardwalk at Iowa Avenue. But like so much else in the early days of the casino gambling era, those plans were scrapped for a bigger, gaudier, and supposedly better design. As a result, a fifty-million-dollar renovation plan that would have seen a casino hotel up and running by 1979 became a three-hundred-and-thirty-million-dollar development project that wasn't completed until the end of November 1981.

The Trop, a twenty-story pyramid-like building, was the epitome of 1980s glitz and glitter. Among its amenities were lavish restaurants like Il Verdi and Le Paris, an eighteen-hundred-seat theater billed as bigger and better than Radio City Music Hall, a rooftop swimming pool and tennis courts, glass-enclosed elevators with a view of the ocean, and 521 guest rooms, at least half of which were equipped with the latest in bathroom chic circa 1981—Jacuzzi-style tubs. But no one had any illusions about the hotel's real attraction: its 1,389 gleaming new slot machines and 116 gaming tables.

Previte saw right away that the rush to complete the job offered him the opportunity to make a few scores. His first security assignment was at a huge warehouse just outside of Atlantic City where Ramada was storing most of its equipment and a lot of its food. Everything for the Tropicana came through the massive staging area—furnishings for the hotel rooms, linens and glasses for the bars, silverware and crystal for the restaurants, seats and sound equipment for the concert halls and theaters.

Previte estimates that about a million dollars worth of equipment and furnishings went out the back door during his watch. One friend, he said, opened a restaurant in the Hammonton area and stocked it completely with Tropicana "surplus" Previte sold him. The chairs and tables, the napkins and linens, the glasses and silverware—all of it came from the warehouse. There was so much inventory, so much activity, and such a rush to get the casino opened that nobody was able to keep track of it all.

"I used to take five or six boxes of frozen lobster, fifty-pound boxes, and sell them to local restaurants," he says. "That was another paycheck a week for me. There was just so many ways to make money."

A short time after the Tropicana opened, Previte moved to a security detail, working both the casino floor and the hotel. He smiles when he talks about those days. The casino, he says, was "an open air-market for crime."

He was a facilitator.

"This mob guy from Newark, Bobby Cabert [Robert Bisaccia, a Gambino family capo since jailed on racketeering charges], used to come down all the time," Previte said. "I'd get him a room, get him meals, get him broads, whatever he wanted. He always had a good time. We got to know one another pretty well."

At one point, Cabert suggested that Previte might want to go to work for him.

"You couldn't pay me what I'm stealing from this joint," Previte replied.

Cabert laughed. "You're probably right," he said.

Previte would supply casino patrons, particularly high rollers, with the extras that weren't included on the casino comp list. Hookers, cocaine, marijuana, entry to a poker game, a bookie who could take action on any sporting event: these were some of the "legitimate" services he provided—always for a fee, of course. He also made a score whenever he was assigned to "escort" a drunken patron back to his room. The next morning the gambler would wake up with a headache—and substantially less in cash and chips than he had when he left the gaming tables.

There were also more imaginative scams.

"I had a desk captain I used to work with," he said. "Sometimes a high roller would come and get comped for four or five nights, but might leave after the second night. We would sell the room."

Previte had a list of customers who were ready to buy in. The desk captain would put them in the empty suite, using the previous guest's comp for meals, drinks, whatever they could get, until the comp expired. They might pay Previte and the desk captain five hundred dollars for a room that cost twice as much, then run up food and drink bills that far exceeded even that.

Nobody cared. It was a comp—a tax deduction for the Trop, a virtual freebie for the phony high roller, and another score for Previte.

Sometimes Previte's scams were less subtle.

"We used to rob the safes where customers would store their cash and jewelry," he says. "I'd make sure the security camera was pointed in another direction, then one of the desk people who had access would go back and grab what he could. We'd fence the stuff. Most times, the casino would make good to the patron. They'd rather pay the money than have a story come out about a safe being robbed."

With each new casino that opened, the competition for the big players grew more intense. None of the casinos could afford a whispering campaign—and there was always a competitor out there ready to stab another casino in the back and grab a customer away.

Previte understood this implicitly. It was his world.

On most nights he'd have three or four rooms set aside, empty rooms for which he had the keys. He'd installed hookers in the rooms, and he and several associates would steer customers their way, generating thousands of dollars in income each night.

They also discouraged competition.

"There were a lot of escort services that would send girls up," he said. "I'd put a guy in a room and have him call for a girl. When she arrived on the floor, I'd bust her. She'd say she was there to see so-and-so. We'd knock on the door and he'd say he didn't know anything about it.

"Now she's trespassing, so I threaten to have her arrested, have her picture taken, all things that are bad for her business. Then we negotiate a deal." The price was usually paid out in money and sex. The girl would give up a little of each in order to walk out of the casino without any legal problems.

For five years Previte worked the security detail at the Tropicana. He was making a salary of between fifteen and twenty grand. He had a wife and two young daughters and a house with a mortgage in Hammonton.

He also had a life as a high roller.

"Atlantic City wrecked my marriage," he now says. "No doubt about that. I was never home. I just had too much going on."

Previte would finish a four-to-midnight shift, then change into civilian clothes and walk out in front of the Tropicana, where a limo from another casino—usually Harrah's—would be waiting for him. He'd be driven to the casino, over by the marina, and gamble into the wee small hours of the morning.

"I'm making twenty grand in salary and I'm gambling that much in one night," he said. "Something's wrong here, but nobody notices. One night I started a roll at four A.M. and didn't seven-out till like five-thirty. I won sixty-nine thousand. Another night I lost like forty-two grand. . . . But I'm making all this money. Day and night I got all these things going on.

"To tell you the truth, I got tired. In a way I was glad when it ended."

3

Previite had two concerns while wheeling and dealing at
the Trop.

One was the New Jersey State Police. The other was the
mob. Eventually he would have to deal with both of them.

The mob showed up first.

"I get a call from the front desk one day and they tell me
there's a guy down there who wants to see me," Previte recalled. "He said his name was Leon Phillips."

Previite took the elevator down to the lobby. He approached the front desk and saw a well-dressed, dark-haired
guy waiting. He recognized him immediately.

It was Phil Leonetti.

The young mobster didn't waste any time. "You know
who I am?" he asked.

"Sure, I know," Previte said.

"Then I guess you know why I'm here."

"I can pretty much figure that out," Previte replied.

Leonetti told Previite that he and the people he was with
couldn't allow him to operate without paying them something. It just wouldn't be right. Leonetti said he wanted two
thousand dollars a week.

"I told him there was no way I could pay him two thousand," said Previite. "Although at that time, I probably could
have. But we negotiated. I ended up agreeing to five hundred
a week. I knew I hadda pay. It was the cost of doin' business.
I didn't want no problems."

After that, Leonetti would come once a week, sometimes every other week. They would meet in the lobby, then walk over to a spot near the indoor parking garage where Leonetti knew there were no surveillance cameras. Previte would hand the young wiseguy an envelope. Leonetti would smile, thank him politely, and walk away.

"You know, he said he was there for his uncle, Nicky Scarfo," Previte said. "But I was never sure about that. I never met Scarfo. I knew who he was; everybody knew who he was. But I just had the feeling that Scarfo didn't know about me. I think Leonetti just put that money in his own pocket. But I didn't care. It was something I hadda do. And at that point five hundred a week was no big deal. I was making five, six times that a night, sometimes even more.

"There was just so much money, if you were smart and if you weren't afraid to take chances."

Like Leonetti, the New Jersey State Police also knew Previte was doing serious business. But it would take them much longer to confront him. Previte figures he was paying Leonetti for about two years while bobbing and weaving around a state police investigation that had his name on it.

"I'd get a call from my guy on the front desk telling me to be careful," Previte said. "He'd say there were some undercover people working the casino. They used to bring detectives in from other parts of the state. They knew I was doing stuff, they just couldn't prove it."

One night, after Previte had gotten a tip about an undercover sting that was about to go down, a woman claiming to be from an escort service showed up on the floor where he was working. Previte's antennae went up.

"I arrest her for trespassing," he said. "She goes, 'Can't we work something out?' I say, 'I don't know what you're talking about.' I bring her downstairs and I insist that we file a complaint, strictly by the book. My supervisor gets all flustered. He knows about the sting, but he can't say anything. 'Let's give her a break,' he says. 'No way,' I tell him. I was just breakin' his balls."

A few months later, Previte got another tip about an undercover operation. On that night he was carrying keys to four hotel rooms where hookers were working, he had a stack of football pools and cash he had collected, and he was holding a small quantity of cocaine and marijuana for a customer who was planning a party that night with one of the girls.

"I'm coming down the elevator and I got a bad feeling," Previte said. "I dump three of the keys, but I forgot about the fourth and about the drugs. It wasn't really that much. I also got rid of the football pools."

Previte walked off the elevator and into the arms of two undercover state police detectives. They escorted him to his supervisor's office, where they told him the party was over. He was charged with theft of services and drug possession. Handcuffed to a chair, he was questioned by both the state police investigators and some officials from the Trop.

"We know what you've been doing," one arrogantly told him.

Previte just rolled his eyes.

"You've taken a million dollars out of here," said the executive.

Better multiply that by four or five, Previte thought.

"Look, the charges didn't bother me," Previte said years later as he described the incident. "Theft of services because I had the key to a room; a small amount of drugs. These weren't serious. It was the other stuff that could have been a problem—but they couldn't prove any of that. I knew it and they knew it."

One of the detectives, John Sheeran, offered a deal. "We can make these charges go away," Sheeran said, "if you talk to us. Give us something we can use."

Previte hesitated, but eventually he decided to listen. He knew Sheeran by reputation. He knew the stocky state police

detective was everything he had never been as a law enforcement officer; Sheeran was totally honest and completely incorruptible. Previte knew he could trust him.

Previte's decision to cooperate, he says, had nothing to do
with being a rat, or beating the charges. It was about doing
the unconventional. Previte had survived and prospered for
more than twenty years by being bold and by taking risks.
This was another gambit, another scheme, and one that
under the right circumstances might provide a substantial
payoff. He knew he'd get off with a fine and probation if the
case were pursued. He'd lose his job, but that was inevitable
now that he had been arrested.

This was bigger than that. Like a chess player, he was
thinking ten or twelve moves ahead. His decision, back at
the end of 1985, to cut a deal with the New Jersey State Police set him on a course that would eventually lead to a
major role in the decimation of the Philadelphia mob. It
would also guarantee that Previte, unlike the dozens of other
mobsters with whom he eventually did business, would
never spend any time in jail.

At the time Previte was considering these options, the mob
was already in self-destruct mode.

Nicky Scarfo had become an underworld despot—paranoid, violent, and motivated primarily by his desire for cash.
Like John Gotti, who took over the Gambino family after or-
chestrating the December 1985 assassination of boss Paul
Castellano, he added a new level of arrogance to an already
dangerous mix of incompetence and greed. The two mob
bosses, one in New York, the other in Atlantic City, were
Mafia bookends. They epitomized La Cosa Nostra in the
1980s. In many ways they were the underworld reflection of
the pirates who were running rampant on Wall Street at the
time, guys like Michael Milken and Ivan Boesky, entrepre-
neurs who based all their business on their own personal
bottom line. It was all part of the go-go eighties, the era of

Oliver Stone's *Wall Street*, of Gordon Gekko's advice to his young associates, "Greed is good."

Scarfo understood the sentiment perfectly.

Angelo Bruno had used murder as the negotiating tool of last resort; for Scarfo it became a calling card. A dozen mobsters were killed as he consolidated power and extended his control over an organization that should have been poised to expand and legitimize itself with the coming of casino gambling. Instead, the Philadelphia crime family took a step back, returning to a bloody era of internecine power struggles and wanton violence. Scarfo could have been a latter-day Frank Costello, well-heeled and well-connected. Instead he fashioned himself as the new Albert Anastasia, the notorious and homicidal leader of Murder, Inc.

Previte, who followed these developments with the interest of someone who traveled in that world but did not belong to it, saw his deal with the New Jersey State Police as an insurance policy, "something I could keep in my toolbox and pull out when I needed it." The deal was straightforward: he would give up his job at the Trop, and in return the charges against him would be dropped. He and an associate, a guy named Freddy Aldrich, would become deep background informants for the state police, supplying investigators with information—but not testimony—in exchange for a weekly retainer of three hundred and fifty dollars.

"I wanted Freddy involved because he knew a lot of people in the drug business," Previte said. "And at first that's what we gave them, a lot of drug cases. We just went about our business. They told me when I made the deal that I would have—what's the word—impunity? They said as long as I didn't kill anybody or beat anybody into bad health, I would be okay."

What the state police didn't know was that Aldrich had to kick back a hundred and fifty dollars to Previte each week.

"I set the deal up," Previte said. "I figured that was only fair."

For the next decade, Previte was a deep background law enforcement informant, first for the state police and later for the FBI. He took cash on a weekly or monthly basis in exchange for providing information about drug dealers, loan sharks, auto theft rings, and, eventually, mobsters. At the same time, he moved his own power base from the Tropicana—where he was persona non grata—to Hammonton. And eventually he moved up the mob ladder, from associate to made member to, finally, capo.

In the old days, Previte knew, that would not have been possible. His status as a former cop would have eliminated him automatically from consideration for any position other than underworld associate. But the rules of the game were changing. And so were the attitudes of those running the organization.

"I brought them money," Previte said succinctly. "So they overlooked everything else."

Before he became a mobster, though, Previte had six years to build a substantial financial portfolio. He was a freelance gangster and racketeer operating with a pass from the state police. And he made it pay off—big time.

Divorced now, and living in a sprawling brick ranch-style home just outside Hammonton, Previte had his own highly efficient and tight-knit crime family, a group of six or seven loyalists who would do anything for him. He had connections with both the state and local police. In fact, a Hammonton police detective—the son of the local police chief—was a friend and business associate. And he had money invested in both legitimate and illegitimate business operations, ranging from a pizzeria and a restaurant to a sports book and a whorehouse.

"There were people who were specialists," Previte says. "A guy runs a book, lends money, does burglaries. I was like a general practitioner. I did it all."

Previte's last legitimate job before becoming a full-time underworld entrepreneur was as an inspector for the New

Jersey Racing Commission. In the late 1980s, a year or two after he left the Trop, he worked at the Garden State Racetrack in Cherry Hill and the Atlantic City Racetrack in Mays Landing.

He had a close friend line up the job for him. On paper, he was the perfect candidate. To someone looking to hire an experienced security person, his record would have looked clean, even exemplary. He had an honorable discharge from the Air Force. He had retired on a partial disability from the Philadelphia Police Department. And he had resigned from his job as a security guard at the casino.

None of his bad acts had followed him.

At the racetracks, one of his jobs was to take urine samples from winning horses after races. These were then tested to make sure the horses hadn't been doped—given drugs to enhance their performance.

Previte quickly found a way to turn that seemingly demeaning job into a moneymaker. Some friends of his had connections to a group of fixers, guys who would routinely rig a race and then bet heavily. Previte was promised cash and information—that is, tips on the right horses to bet on. In return he would make sure that the horses came up drug-free after the races.

"The day before a race, I would collect urine from horses that I knew were clean," he said. "I'd keep it in my car. Then on race day, I'd substitute the clean piss for the piss I got from the doctored horse."

There was one problem. Fresh urine was warm; urine that had been sitting for a day was not. But Previte figured a way around that. "I had a small microwave oven rigged so that I could plug it into the lighter in my car. I'd heat up the good urine in the microwave before I turned it in."

Over two seasons Previte helped fix a couple of dozen races, pocketing over a hundred thousand dollars.

In his other business ventures, of course, things didn't always go as smoothly. There were gamblers who didn't want to pay. There were debtors who would borrow ten or twenty grand and then disappear. And there were friends—friends of his, friends of his associates, guys they had grown up with, guys they had gone to school with or played football with or who had married one of their cousins, guys who always had an excuse for not coming up with the money that was due.

"I used to hear these sob stories all the time," Previte said. "You can't listen. You can't feel sorry for a guy. He'll tell you his kids are sick, his wife's in the hospital, he's three months overdue on his mortgage and he's gonna lose his house. My guys would come back with all these stories. They'd be feeling guilty about asking for the money. I'd have to tell them, 'When this guy was placing the bet, was he worried about his kids? Was he talking about his wife when he borrowed the money? Was he concerned about his mortgage?' This was a business. This wasn't a charity.

"Then I'd have to take somebody with me and go make the collection myself."

Fear and intimidation, an occasional beating—that was his style.

"I spent my whole life trying not to kill people," he said. "I had a good thing, and that would ruin it. . . . Murder is bad for business. You create problems for yourself, plus you don't get paid. How's a dead guy gonna pay ya? . . . Don't get me wrong, I would have killed if I had to. There were times when I would have liked to have killed somebody, a guy would get me that angry. And there were times when I beat a guy into bad health. But no murders."

The fear of being murdered, or of being badly beaten, was often more effective than the act itself. Previte knew that.

"I had a better time intimidating people," he said. "And I usually got my money."

There was the time he jammed a .38-caliber pistol into the mouth of a deadbeat gambler who owed him twenty thousand dollars. The guy happened to be a prominent professional football player who thought he could walk away from his financial losses because of his status.

"Of course I took the bullets out beforehand," he said. "I didn't want to have an accident. Sometimes in that situation, a guy will jump and the gun could go off."

The hapless victim, however, thought the gun was loaded. "He threw up," Previte said with a laugh. "I guess I put the gun too far down his throat."

There was also the time when he and two associates grabbed another deadbeat—this over a ten-thousand-dollar debt—and terrorized him by sticking his head under the hood of a late-model Lincoln Continental.

"We had his head like two inches from the engine fan," Previte said. "We had it jammed in there. Then we'd rev the engine."

Once, twice, three times they went through the routine, laughing, cackling, and threatening as they brought the victim up for air, then shoved his head back under the hood where he could feel the fan, smell the engine oil, and breathe in the smoky soot and fumes.

In both instances, Previte said, the gamblers had pleaded poverty, claiming they didn't have the cash to satisfy their debts. Within two days of each confrontation, Previte said quietly, they came up with the money.

"Things were good," Previte said. "I was making a lot of money. Nobody was bothering us. It might have been the best time I ever had. I would guess I was making about a million a year. I kept a half million and my guys split the other half million. I had money I didn't know what to do with, big money. See, I wasn't that well known at the time. That was good. Nobody was paying attention to what I was doing. Later, after I got Mafia-ized, I had a lot of eyes on me."

Scarfo, meanwhile, was running the Philadelphia mob into the ground with his arrogant, shoot-first-and-ask-questions-later approach.

As a result, when two of his top associates, Thomas "Tommy Del" DelGiorno and Nicholas "Nicky Crow" Caramandi, found themselves jammed up, they quickly realized they had nowhere to turn but the FBI.

DelGiorno, a capo in the organization, had been picked up on a New Jersey State Police bug complaining about Scarfo's management style. Among other things, he called his boss an incompetent and claimed that "four Irish guys from the Northeast [of Philadelphia] could run a better organization." Caramandi got caught in a federal sting operation while trying to shake down a developer who had a multimillion-dollar contract to develop the Philadelphia waterfront.

Both mobsters had screwed up colossally. Both knew how Scarfo dealt with those he no longer trusted.

"There was no talking to Scarfo," Caramandi said. "Once he made up his mind, you were as good as dead."

Both DelGiorno and Caramandi cut deals with the feds, agreeing to cooperate. Their testimony in a series of gangland trials in the late 1980s destroyed the Scarfo organization. The subsequent convictions also led to at least three other major defections, including Scarfo's nephew and underboss, Philip Leonetti. The same "Leon Phillips" who had shaken Previte down decided in 1989 to cooperate with the government.

Previte smiled. He and Leonetti were back on the same side of the fence. But it would be years before anyone else in the underworld knew it—and by then it would be too late to do anything about it.

Nicky Scarfo was convicted in 1987 of conspiracy in connection with the attempted waterfront development extortion. A year later he and sixteen other members of the organization were found guilty of racketeering charges, in a sweeping organized crime case that included nine murders, four attempted murders, and various other counts of loansharking, gambling, extortion, and drug dealing.

Little Nicky was sentenced to fourteen years in the conspiracy case and got fifty-five years for the racketeering conviction. The sentences were set to run consecutively. Scarfo
was fifty-eight at the time of his first conviction. The combined sixty-nine-year prison sentence was the equivalent of
a life sentence. Like Gotti, he will probably die in jail.

A year after Scarfo's second conviction, John Stanfa was
released from prison. He had served nearly seven years of an
eight-year sentence for lying to the grand jury investigating
the Bruno murder. Originally, according to underworld
sources, Stanfa was supposed to return to his native Sicily.
That was the agreement Gotti and Scarfo had worked out.
Stanfa had used his Sicilian Mafia pedigree—he came from
the town of Cacamo near Palermo; several family members
there were major Mafiosi—and his family ties to the Gambino organization to get a pass on the Bruno murder in exchange for leaving the country.

But with Scarfo gone, that deal was sweetened. Stanfa
was encouraged to return to Philadelphia to put the troubled
mob family there back together. Eventually he reached out
to Previte.

"He sent a couple of guys down to see us," Previte said,
"but they were fresh. Especially this one kid, Phil Colletti.
He starts acting like a tough guy, telling me we gotta pay
'em this and give 'em that. We told them to go fuck themselves and sent them back to Philadelphia. The next thing I
know, I get a call that Stanfa wants to see me."

They met outside the Melrose Diner in the heart of South
Philadelphia. At the time, Stanfa was living a few blocks
away in a row house on Passyunk Avenue. He had been out
of prison for about a year and was struggling to reorganize
the crime family. He had a group of older guys who had survived both the Scarfo violence and the federal prosecutions.
And he had a group of younger guys—the sons, brothers,
and nephews of Scarfo family members—South Philadelphia street-corner kids who thought their spots in the organ-

ization were a birthright. Joey Merlino and Michael
Ciancaglini headed that group.

"Let's walk," Stanfa said to Previte.

"He wasn't a bad guy," Previte recalled. "He had his good
points and his bad points. But I liked him. He was old-
fashioned. An old-fashioned Siggie. I understood him."

At first Stanfa wanted to know why Colletti had been mis-
treated.

Previte decided to be blunt. "You sent a punk," he told
Stanfa. "He came down there woofing. He wasn't a gentleman."

Stanfa nodded. He understood.

"I heard good things about you," he said finally. "Maybe
you could help me out."

This was early in 1991.

By 1993 Previte was part of Stanfa's inner circle. By that
point, the mob boss had moved with his family to a sprawling ranch-style house on five acres in Medford, New Jersey.
In the midst of a mob war, Stanfa considered the suburbs
more secure than his former South Philadelphia home. Each
morning either Previte or Aldrich would pull up in Previte's
silver Cadillac Seville, pick up the mob boss and his son Joe,
and drive them over to their food distribution warehouse in
South Philadelphia.

On the morning of August 31, 1993—three weeks after
the ambush in which Mike Ciancaglini had been killed and
Joey Merlino had been wounded—Freddy was driving.

4

"I was supposed to drive that morning, but I had a doctor's appointment, so Freddy picked up John and his kid," Previte said. "Afterwards, John was a little suspicious. He wondered if I had anything to do with what happened. He mentioned it to Freddy, not directly but just like 'Where was Ronnie?' Right away I knew what he was thinking, so I talked with him about it. He said he never really had any doubts about me, but I knew from the question he asked Freddy that he was wondering.

"See, John always underestimated the kids. He figured after Mike Ciancaglini got it, it was all over. You gotta give them credit. They had a lot of balls coming back like they did."

Balls might be one way to describe what happened that morning. Insanity would be another. What was clear was that Joey Merlino did not intend to go quietly.

In the middle of rush-hour traffic on the Schuylkill Expressway, an eight-lane highway that brought thousands of commuters into Philadelphia each morning, two gunmen in an unmarked white van opened fire on John Stanfa as he was being driven to work.

The shooting occurred a little before 8 A.M. Stanfa, riding in the front passenger seat, had the presence of mind to duck when the gunmen, firing from two makeshift portholes that had been cut in the side of the white Chevy Astro, began strafing the side of the car with 9-millimeter machine pistols. Stanfa's twenty-three-year-old son, Joe, riding in the backseat, was not as quick. He took a bullet in the cheek.

Aldrich, a decorated Marine veteran who had served two tours of duty in Vietnam, may have saved the lives of both Stanfas that morning. When the shooting started, he rammed the Seville into the side of the van, denying the gunmen the angle they needed to continue firing at their targets. Aldrich forced the van off the highway, then gunned the engine of the Caddy and headed for the next exit, about a block from Stanfa's food distribution warehouse. With one tire shot out and the other burning rubber, the bullet-riddled Cadillac lurched to a stop in front of the business. Joe Stanfa was carried to another car and rushed to the emergency room at the Hospital of the University of Pennsylvania, about a mile away.

The shooting was more Sicily than South Philadelphia. And that may have been the point. This was a message delivered in a language Stanfa clearly understood. Police were amazed that no one else had been hurt. Hundreds of motorists who were put at risk arrived at work that morning to hear on the radio about the attempted assassination of the city's mob boss. Many didn't even realize they had been in harm's way. One stray bullet crashed through the bathroom window of a small apartment in a housing project adjacent to the highway.

Nothing else was hit.

The day after the shooting, the van was found by police, parked on a street a few blocks from the highway. It had been reported stolen two months earlier from a parking lot near Veterans Stadium.

Detectives noted that the interior panel wall of the Astro was pockmarked with bullet holes, indicating that the gunmen had apparently been firing wildly from inside the vehicle, and not always through the portholes.

The incident—a new and more sinister Philadelphia version of *The Gang That Couldn't Shoot Straight*—reestab-

lished Joey Merlino as a presence in the underworld, one
that neither Stanfa nor law enforcement could take lightly.

Joe Stanfa remained in critical condition for more than a
week. But he would recover.

The mob war that many thought had ended abruptly three
weeks earlier, with the murder of Mike Ciancaglini, was actually just beginning. Now everyone was fair game. Stanfa
wanted revenge.

In the midst of it all, the feds recorded a series of prison
phone calls Ralph Natale made to Merlino. If Skinny Joey
was worried about retribution, or concerned about investigators linking him to the shooting on the Schuylkill, he wasn't
showing it. On the calls he talked openly about gambling and
betting and problems he was having with his landlord. The
landlord, concerned over the shootings, had suggested that
Skinny Joey and his girlfriend find a new place to live.

"He thinks I'm gonna get killed," Merlino told Natale. "I
said, 'Fuck you.' . . . I said, 'You got my first security deposit
and my last security deposit and I pay my rent.' . . . I'm
pretty sure the guy don't want me in the building no more."

"Cuz, you're gonna live forever," Natale said with a
laugh.

In another conversation Merlino agonized over the fact
that Stevie Mazzone, a top associate, had failed to play his
young daughter's birth date as the daily number. It had come
out "straight as an arrow," Merlino said. Mazzone's daughter was born on September 1. The number that day was 901.

"I said, 'How the fuck can you not play it?' " Merlino told
Natale. "He was sick."

Merlino, who throughout his career would be described as
a degenerate but not particularly astute gambler, was also
recorded offering a handicapping assessment that any long-
suffering Phillies fan would appreciate.

The 1993 season was a magic year for the Phillies. A team
built around Lenny Dykstra, Darren Daulton, Pete Inca-

viglia, and John Kruk—a team of dirtball, blue-collar ballplayers who knew how to grind it out—rode into the World Series and into the hearts of Phillies fans everywhere. It all came apart, of course, in Toronto, when Joe Carter took Mitch Williams, the Phillies stopper, deep in game six. Nearly two months before that tragic night, Merlino had told Natale, "The Phillies gotta go out and get a closer. They ain't never gonna win with this guy Mitch Williams."

Natale and Merlino were also recorded making oblique references to the mob war.

"You know the cream always comes to the top," Natale said in one of his cheerleading speeches.

Merlino also bragged about how he had refused to talk with Philadelphia police detectives investigating the shooting of Mikey Chang. "I ain't got nothing to say," he said he told the detectives before chasing them off his front step. Both Natale and Merlino, in subsequent phone calls, acted completely in the dark about the expressway ambush, and scoffed at newspaper reports tying Merlino to that shooting.

"Don't let people psych you out, cuz," Natale said.

Over the next six months, the FBI, the state police in New Jersey and Pennsylvania, and the Philadelphia Police Department intensified their pressure on the mob. Any wiseguy from either faction spotted in a vehicle was automatically stopped for a traffic violation—failing to signal when changing a lane, failing to come to a complete stop, driving erratically, it didn't matter. Cars were searched. Weapons were confiscated. Each week, it seemed, there was another report of a mobster being pulled over, then arrested for the illegal possession of a gun.

In most cases, the charges were dismissed. The stops were patently illegal. But the goal was not traffic enforcement, it was a desperate attempt to get as many guns as possible off the streets.

Still the shooting continued.

Two Stanfa associates were ambushed on a street corner,

narrowly avoiding assassination. A Merlino crony was shot and killed in front of a diner. A bomb planted under Merlino's car failed to detonate. A sniper with a rifle and scope failed to get a clear shot at Merlino as he entered a nightclub. A Stanfa associate suspected of cooperating with authorities was shot in the back of the head.

The FBI continued to track the action, using the bugs planted in Stanfa's lawyer's office in Camden. The feds also began to rely more and more on an informant who was positioned to give a firsthand account of the action.

"I figure I saved a half-dozen lives," Previte now says. "And the thing is, I was saving the lives of guys who wanted to kill me."

Not all the routine traffic stops were happenstance. If Previte knew Stanfa had assigned a hit team to carry out a killing, he would call his new FBI handlers. They would pass the word along to the Philadelphia police, and the gunmen would be stopped before reaching their target. Or a team of hit men might arrive at a location and find the police already on the scene.

"I'm out there every day," Previte said. "Merlino's guys want to kill me, and I'm tipping off the FBI about plots to kill them. I musta been nuts. The FBI said, 'Don't worry, we got your back.' Right. I hadda look out for myself."

Previte kept the bullet-riddled Cadillac in his backyard, a daily reminder of the turbulent times in which he was living. Years later he gave the car to Aldrich, who auctioned it off via the Internet to a Wilmington, Delaware, man who was fascinated with the mob. Aldrich sold the car for two grand. "He shoulda got more," Previte said.

"I think I went a year without sleeping in my own bed," Previte said of his life during the Merlino-Stanfa war. "I used to sleep on the floor. I put a dummy in the bed. People knew where I lived. I didn't want to make it easy for anybody.

"Freddy [Aldrich] was living with me for a while back then. He's sleeping in the guest bedroom, also on the floor.

One night I hear him screaming and thrashing around. I go in his room. He's got his arm all tangled up between the mattress and the bedspring. I think he was having a nightmare. He used to get them, about Vietnam. Now he's trapped in the bedspring, screaming and cursing. I wake him up. He looks around, realizes where he is and, he goes, 'Fuck it. If they kill me, they kill me.' And he flopped down on the mattress and went to sleep. After that, he stopped sleeping on the floor."

Previte's move from the New Jersey State Police to the FBI came after he had burned several bridges with New Jersey authorities. The last straw was an extortion attempt in Glenolden, Pennsylvania, in April 1992. But the split between Previte and his New Jersey handlers had been coming for months, if not years.

There was a faction within the state police and the Division of Criminal Justice (the branch of the attorney general's office that prosecuted most mob cases) that had grown tired of Previte's "antics."

"What did he do now?" was the phrase most often uttered when Previte's name came up at their weekly planning meetings. The detectives Previte reported to understood life on the streets. The lawyers and supervisors who worked in the office did not.

Previte, of course, didn't make it easy for those who were in his corner. He enjoyed pushing things to the limit, seeing how much he could get away with. It was his nature. If he were a child, he'd be called incorrigible. He never allowed his role as an informant to diminish his criminal activities. He was out every day committing crimes, dealing with criminals. That's what made him a valuable informant. But it's also what drove the bean counters and deep thinkers crazy.

An associate in Hammonton once tried to explain the difference between Previte and most of the other mobsters who were out on the street at the time. "This is not a sideline to him," the guy said. "This is his life. This is his business. He is Cosa Nostra. . . . This guy knows everything about every crime there is. There ain't nothing he ain't did. . . . He's like a New York guy. He's got so many things going right now. He is crime personified."

A street cop understands the value of someone like Previte. A street cop also has the ability to look the other way. But a prosecutor building a criminal case has to be responsible about the source of any evidence he hopes to put in front of a jury.

Previte was invaluable to the guys in law enforcement who worked the front lines. But he was a nightmare for their supervisors and the people in the attorney general's office. And the fact of the matter is, none of them knew half of what he was doing.

"People were always coming to me with deals," he said. "There was this guy who owned a pizza shop in Pleasantville [a small town just outside of Atlantic City]. I borrowed fifty grand from him, told him I was gonna put it on the street. Loan-sharking. It was a scam. I put the money in my pocket. Couple of months later he calls and says, 'What's up?' I told him we got burned. Lost the money. He didn't care. He was just happy to be with me. He had a guy who owed him four thousand. Asked me to collect. I went to see the guy. Got the money. Kept half. He still didn't care. Used to invite me to come to his restaurant all the time and eat. The spaghetti wasn't even that good. But that's the way it was. People want to be with you."

Not everyone.

Previte had a running battle with the owner of the Silver Coin Diner in Hammonton. The Silver Coin was a popular lunch and dinner spot located on the White Horse Pike, the highway leading to Atlantic City. Previte would go to the diner almost every night, taking the same booth—the second on the left as you entered—and holding court for hours, conducting business, meeting associates, doing deals.

"If you wanted to see Ronnie you went to the Silver Coin," said another friend from that time. "And if you didn't, you didn't go there."

Previte, who despite his size and girth always managed to attract beautiful women, eventually started dating one of the waitresses, a local girl who looked like Natalie Wood. And after she hooked up with Previte, there were those who said she started to act like a movie star as well. She was eventually fired for failing to come to work on time. Depending on who's telling the story, the young woman was let go either because she was too close to Previte or because she thought her connection to Previte gave her a free pass to disregard the rules of her job.

Shortly after the waitress was let go, the front windows of the diner were peppered with a blast from a shotgun one night after closing. On another occasion a fire bomb was found outside the restaurant's back door. The harassment stopped after she was rehired.

Previte just smiles as he recounts the "bad luck" that befell the diner.

The owner knew he couldn't complain to the Hammonton police. At the time, Previte was tight with both the chief and the chief's son, a detective who, according to one FBI report, was taking $250 a week to protect Previte's bookmaking operation. But the owner of the diner had a friend who was the captain of a New Jersey State Police unit based in Atlantic County. The captain, who wasn't part of the organized crime squad, was unaware of Previte's working relationship with law enforcement. As a favor to the diner owner, he confronted Previte in the parking lot of the Silver Coin one night. Words were exchanged, but little else happened. That appeared to be the end of it.

And in most cases, it would have been.

But Previte had to have the last word. One night a few weeks after the confrontation in the diner, he snuck up on the police captain's home, which was located just outside of Hammonton.

"I had a friend who was a crop duster fly over one day, so I knew the lay of the land," he said. "I had to walk through these woods and muddy fields to get to the house. I didn't want to go from the road."

Previte said he crept up to the back of the house and spray painted a happy face and a sad face on the bedroom window where he knew the state police captain and his wife were sleeping.

"Under the happy face I wrote 'ME,'" Previte said. "And under the sad face I wrote 'YOU.'

"I just wanted him to know I knew where he lived and that I could get to him. Hammonton was my town. Nobody was gonna tell me where I could or couldn't go. It was intimidation. He wasn't going to intimidate me. I was going to intimidate him."

Previte continued to hold court at the Silver Coin. In fact, he was a regular there until his role as a cooperating witness became public. But it was that kind of behavior that soured his relationship with the New Jersey authorities. That and his blatant disregard of an order.

"I'm working with Stanfa, right, and he wants me to go over to Delaware County, Glenolden, with a couple of his guys and shake down this bookmaker who wasn't paying us anything," Previte said. "The people with the New Jersey State Police tell me I can't go into Pennsylvania and commit a crime—that if I do that, our deal is over. It was ridiculous. How was I supposed to explain that to Stanfa? So I went."

The bookmaker owned a meat store. Previte went in to "negotiate" while the other two mobsters stayed outside in the car. It was a setup. Cops swarmed into the parking lot within minutes of Previte's arrival. The two Stanfa goons quietly drove away. Previte was busted on charges of conspiracy, making terroristic threats, and theft by extortion.

The charges were later dismissed, but the arrest effectively ended any deal he had with the New Jersey State Police.

"There were guys over there I still talked to, John

Sheeran, Chuck Atkinson, guys I would give information to, but that was because I liked them, not because they were paying me," he said. "Sheeran set me up with the FBI and I worked out a deal with them."

Previte started working with the FBI late in 1992. At first his arrangement was the same as it had been with the New Jersey State Police. He would provide information in exchange for a monthly stipend and he would never have to testify.

In fact, he said, one of the first agents he worked with recommended that he never agree to become a witness. "He knew me," Previte said. "He knew the kind of life I lived. He knew becoming a witness would change all that and that I wouldn't like it. He was right. But things happened.

"It got to the point where I knew I was gonna get burned, one way or another. Either the law would get me, or I'd wind up dead like those other motherfuckers.

"I'm a bit of a mercenary. I never believe in standing pat, staying in one place. If you stay in one place, you're gonna get ambushed. You have to move. Sometimes you have to do the ridiculous or the offbeat to keep other people at bay. I wasn't like a lot of these guys. They get up at twelve, go play cards, eat a cheese sandwich, and go drinking all night. I worked every day. I worked at crime, but I worked.

"I had a great crew where I was. If the mob was really the mob, I'd still be a mob guy. It was such a disillusioning thing to see. It wasn't anything like in the movies or the books. It's nothing like it's portrayed. It's a shame. It was Ali Baba and the Forty Thieves. There was no honor, no loyalty. . . . Most of them guys are dumb people. If they took an IQ test, they'd come out around sixty. I knew I was smarter. I've got an IQ of a hundred and thirty. And I knew the mob was a losing proposition, plain and simple.

"You can't be a real bad guy, making big money, and just have a modicum of intelligence. You gotta be smart. The secret to being a bad guy is, when it's all said and done, you have money. When you walk away, you have a couple of nest eggs."

At the time of the expressway shooting, Previte had been feeding information to the feds for about a year. By that point he had become part of what was referred to in the media as Stanfa's "palace guard." Previte, Aldrich, and Vince Filipelli, a two-hundred-sixty-pound bodybuilder and former Mr. Universe contestant, were always around Stanfa.

Their role was captured on film when Stanfa was subpoenaed to appear before a state grand jury in Trenton in the midst of the war. It was a scene right out of the movies. Stanfa, wearing a neatly tailored silk business suit, shirt, and tie, walked through a gauntlet of television and newspaper cameramen flanked by the three hulking mobsters. Filipelli appeared to be straining under the pressure of a white shirt and tie pulled tight around his bulging neck. Previte, more casually dressed, wore a sports jacket over an open-necked shirt that exposed several thick gold chains. Aldrich, bringing up the rear, was dressed in a windbreaker and cap. Both he and Previte also donned sunglasses for the cameras.

At the time, Previte was labeled a "close associate and confidante" of the mob boss. Later he would be described in court as the "eyes and ears" of the FBI inside the Stanfa crime family. Stanfa trusted him implicitly. He saw Previte as old school, more Sicilian than American. And in many ways Stanfa was right. The mob boss and the FBI informant shared many of the same underworld values. Stanfa, however, believed it was still possible for La Cosa Nostra to function.

Previte was convinced it was over.

In the middle of the war, in an impromptu ceremony devoid of the stylistic ritual that is so much a part of Mafia lore—the pricked trigger finger, the burning holy card, the solemn oath to live and die for La Cosa Nostra—Stanfa initiated Previte into the crime family.

"He told me when the war was over we'd have a formal

ceremony, but he said I was a made guy, a soldier," Previte explained.

It would be one of several extraordinary makings that led mobsters from outside the Philadelphia area to question the bona fides of several members of the organization. In fact, the legitimacy of the family hierarchy was an issue throughout the Stanfa-Merlino-Natale era.

Natale, from prison, kept insisting that Stanfa had no right to run an American crime family. Even before the hostilities began, he told Merlino, Mike Ciancaglini, and the other young wiseguys that the New York families did not recognize Stanfa as boss. Whether Merlino and Ciancaglini needed that assurance to move against Stanfa is another matter. The Merlino organization was more a South Philadelphia street-corner gang than a Mafia crew. It didn't stand on protocol. But Natale, always looking to legitimize himself and play the role of Mafia statesman, insisted that the "rules" of La Cosa Nostra stipulated that a boss had to be born in America. Stanfa, the "greaser," had no right to head the family, Natale explained; he was a poseur. New York would support the move by Merlino and his group to oust the Sicilian-born mob boss.

Natale even claimed to have raised the issue with other mob leaders while he was in jail. He said he got a message to Gotti asking whether the Gambino family recognized Stanfa as the boss of the Philadelphia family. Gotti's reply, Natale claimed, said it all.

"Who knows him?"

After he got out of prison in 1994 and took control of the organization, of course, Natale didn't seem quite as concerned about the rules. When asked who had initiated him into the family, Natale said Merlino had. After he was made, Natale said, he declared himself boss and Merlino underboss.

Whether Merlino, a soldier in the Stanfa family, had the underworld authority to initiate anyone into the Mafia is debatable. And a newly minted soldier like Natale simply declaring himself boss certainly stretches traditional mob rules. But that was what La Cosa Nostra in Philadelphia had become.

It didn't really matter, Natale said arrogantly when asked about this several years later.

"We would have made ourselves if we had to," he said. "Anybody with the cojones could do it."

The Stanfa-Merlino war ended not with the sound of gunfire, but with the FBI knocking on doors.

In November 1993, Merlino, who had been convicted of an armored truck robbery a few years earlier, was jailed on a parole violation. Cited for consorting with known felons—several members of his crew had convictions—he was taken off the streets and given a safe haven at a time when Stanfa gunmen were still looking to cut him down. Then four months later, on March 17, 1994, Stanfa and twenty top associates were arrested on racketeering charges. The multicount indictment unsealed that Saint Patrick's Day morning included numerous counts of murder and attempted murder, including the slaying of Michael Ciancaglini.

The two gunmen involved in that hit, John Veasey and Phil Colletti, were cooperating with authorities. In addition, the feds had hours upon hours of tapes—more than two thousand conversations—from the bugs planted in Stanfa's lawyer's office. Stanfa faced life in prison if convicted. And a conviction seemed almost certain.

Seven months after Stanfa was arrested, Ralph Natale was released from prison. Merlino came home two months later. Together they set about reorganizing the crime family, cleaning up the loose ends from the Stanfa war, and reestablishing order in the underworld.

Because they were both on parole, the two mobsters could not be seen meeting together in public. Natale, who lived with his wife, Lucia, in a penthouse apartment on the Cooper River in Pennsauken, New Jersey, was also prohibited from traveling to Philadelphia, which was about five miles away. He had to stay in New Jersey unless granted specific permission by his parole officer. He and Merlino communicated via intermediaries, Merlino confidantes like Steve Mazzone and George Borgesi.

Natale eventually arranged a job for himself as a "salesman" for a seafood company based in Gloucester City, New Jersey. His "route" included several bars and restaurants in South Philadelphia, and having met with his parole officer's blessing, he made regular trips to the city, where he met face-to-face with members of the organization. Natale and his associates eventually busted out the seafood company, reducing the former owner to a truck driver before sucking all possible revenue out of the business.

By that point, Natale was moving on to bigger and better things. He had been away for nearly fifteen years. He was fifty-nine years old. And he wanted to make up for lost time.

Merlino, with his coterie of young associates, was happy to let Natale assume the mantle of mob boss. Despite what many in law enforcement thought of him, Merlino was no dummy. He may have been only thirty-two, but he had been through a lot. He had seen mob bosses from Angelo Bruno through Stanfa try to run the crime family. Bruno and Phil Testa were both brutally murdered. Scarfo was going to spend the rest of his life in jail. Now Stanfa appeared headed in that same direction.

So Joey happily stepped aside, allowing Ralph to take all the heat and draw all the attention that came with being a mob boss. After he was released from prison, Merlino opened the Avenue Café, a small coffee shop and cigar store on Passyunk Avenue in the heart of South Philadelphia. He and his associates hung there, going about their business, cautiously eyeing the FBI and paying lip service to the boss over in New Jersey.

Ron Previte, at that point, was in no-man's-land.

In fact, more than a few people in underworld circles

began to wonder why he hadn't been arrested in the massive and highly publicized FBI roundup of Stanfa and most of his top associates. Years later, after his cooperation became common knowledge, everyone said that they always knew, that they never trusted the hulking mobster. It was obvious, they said: Previte was not arrested because the FBI wanted him out on the streets. He was a rat.

"If it was so obvious," Previte now asks, "then why did they do business with me?"

The answer to that is somewhat complicated—but as with everything else that had to do with the Merlino-Natale organization, money was a big part of the equation. Previte was an earner, a big-time earner. And as long as he was willing to share those earnings with Natale, Natale wanted him alive.

After Stanfa was arrested, and before Natale and Merlino emerged from jail, Previte hunkered down in Hammonton, continuing his criminal activities and routinely sending money to the prison commissary for Stanfa and several of the others, like Filipelli, who were awaiting trial. It was his way of maintaining his cover: he was doing what a loyal soldier would do.

"I knew I had a problem with Joey and his guys," Previte said. "I was on the other side in that war. But I also knew they weren't gonna come down to Hammonton looking for me. I'da blown them outta the water. So I'm making good money and nobody's bothering me. This is late in 1994, I guess. I get a call one night from this guy Andy Chalaka I know, who owns a restaurant, the Fireside Lounge, on the White Horse Pike in Winslow Township, just outside of Hammonton. He says, 'Why don't you come over for dinner tonight? RN's gonna be here. He would like to meet you.' "

Previte had been expecting the call.

"Besides everything else, it's just a real pain in the ass to deal with a boss," Previte said. "I'd rather just be left alone, but I know that ain't gonna happen. Plus, I don't know what kind of reception I'm gonna get."

Previte called his FBI handler at the time, a veteran agent named Gary Langan.

"I liked Gary a lot. He was no bullshit. And he was a genuine tough guy. He wasn't afraid of anybody. So I call him up and I say, 'Gary, not for nothing, but Ralph wants to meet me tonight.' He says that's great. I says, 'Well, what if I get whacked?' And he goes, 'Well, at least then we'll know who did it.' I'm laughing, but he's laughing even harder. That's the way he was. Then he tells me don't worry, they'll have the place covered. I wasn't taking any chances. I had two of my guys outside in the parking lot with shotguns. I told them if I get carried out, start blasting. I don't tell my guys the FBI's gonna be there, because I don't believe they *are* gonna be there. But I go to the meeting."

Previte drove to the restaurant that night with Anthony Viesti, a friend of his and of Natale's. When they arrived Natale was already sitting at the bar, along with Steve Mazzone, one of Merlino's top associates, and Ray Rubeo, a well-known meth dealer.

Natale was short and stocky, well built and clearly in good shape for a man his age. He had begun a workout regimen in prison, and maintained it after coming home. He would jog, sometimes three or four miles a day, around the Cooper River near his penthouse apartment.

With his shaved head, dark, piercing eyes, and tight-fitting but stylish pullover, Natale looked like an underworld version of Mister Clean, the caricatured muscleman in the once popular detergent commercial. He seemed genuinely happy to meet Previte.

"He stood up, gave me a hug and a kiss. A real gangster, you know. And he introduces me to Stevie [Mazzone]. We shake hands. Stevie says, 'My pleasure,' but real nasty. Viesti's waiting down the other end of the bar. Ralph says, 'Let's talk.'

"He tells me he knows I'm a made guy, that I was just doing my job, and that he wants me to be with him now. I said, 'Ralph, I hope you're not going to ask me to go kill Stanfa's kid, because John was always good to me and I wouldn't do that.' "

Natale told Previte that wasn't his style. Loyalty was important, Natale told Previte that night. He said he knew Previte was loyal to Stanfa when he was with him. Now he wanted Previte to be loyal to him.

"I want you to be my guy down here," Natale said. "I know you're an earner. I heard you got some good guys with you. You're a real gangster, a real gangster. These are gonna be good times. Not like before."

Natale told Previte that some people in the organization—Previte knew he was referring to Merlino—wanted Ronnie dead, but that wouldn't happen.

"It's easy to kill a man," Natale said later, in his best Godfather imitation, while explaining why he decided not to go along with Merlino's request to kill Previte. "It's hard to save one. I figured this guy might have had some talent. We could have made some money with him. He didn't look like a danger to me or Joey or any of the fellows. . . . He had a gambling operation going in the Hammonton area. He had a loan-sharking business going on, and he was able, he said, to dispose of any kind of drugs that we could provide him with."

That night, Previte and Natale talked and drank for nearly three hours. Some of what Natale said made sense. Things had to be run differently. People had to know and trust one another. When Stanfa was in charge, "you guys were like a bunch of gypsies," Natale said. "Nobody knew anybody. Nobody trusted anybody."

Previte nodded and listened.

"He acted like he was so smart," Previte said, "but the whole time I'm thinking, 'You're so stupid to deal with me.' I wouldn't have dealt with me."

In fact, there *were* guys who wouldn't deal with him. Previte, in a twisted, underworld kind of way, respected them for it.

Natale used to hold court nearly every day at a restaurant on the third floor of the Garden State Racetrack in Cherry Hill. He called it his "office." After being recruited by Natale, Previte would go there once or twice a week to check in. He usually brought an envelope stuffed with cash. So did half a dozen other wiseguys who routinely showed up. The FBI, which had the place under surveillance, noted in one of its reports that "Natale usually meets with various members and associates . . . on Wednesdays and Thursdays at the Currier & Ives Room to collect his percentage of the previous weekend's illegal bookmaking profits."

On several occasions when Previte arrived, the group sitting at the bar—betting on the horses, watching the off-track races, or just bullshitting with one another—included former Pennsylvania state senator Henry "Buddy" Cianfrani.

Buddy Cianfrani had once been a major political power broker in the state. But a conviction for running a ghost payroll scam, and other acts of political corruption cost him his senate seat and several years in prison. Still, he remained a player behind the scenes. A South Philadelphian to the core, he also was considered a stand-up guy, never complaining about his conviction and never once offering to cooperate with authorities in order to get out from under his own problems.

The consensus was that Buddy could have sent a lot of politicos to jail if he had wanted to. But it just wasn't in his nature. Previte respected that. He believed that each man had to do what he had to do given the circumstances in which he found himself.

In any event, Previte was convinced that Cianfrani knew he was a cooperator.

"He treated me like I had leprosy," he said. "Every time I showed up at the racetrack, he'd leave. He never wanted to talk around me. He'd shake my hand, say hello. Nice. Sociable. And then in like ten seconds, he'd get up and leave. Somehow he knew. Now that was a smart guy. And I never felt no animosity toward him because of that. You never get mad at a man for being smart."

Natale, on the other hand, wasn't so smart.

"I knew being around Ralph was gonna cost me money," Previte explained. "That's what that first meeting was all about. I started out giving him one G a week. He'd say, 'You don't have to do that,' but I knew that's what he wanted. Ralph was the type of guy you petted. You call him Godfather and you pet him very lightly. I knew . . . those other guys couldn't fuck with me if I was with him.

"Some people might say, 'You rotten motherfucker.' But you have to understand, I had no loyalty to them. They wanted to kill me. To me, they were just marks. Just like I was to them. I don't feel bad about what I did to them.

"You know, the secret to being a smart guy is realizing just how smart you are and what your limitations are. These guys never realized their limitations. . . . I had no respect for them. I knew they were dumb and I knew they were easy."

But they were also ruthless. And they could be deadly.

5

Billy Veasey was on his way to get the bagels when they shot him.

He had just pulled away from his row house in the 2600 block of South Bouvier Street, relaxed and at ease behind the wheel of his new GMC Jimmy. Veasey owned a small construction company; depending on who you talked to, he may also have been involved in the rackets.

For a time, according to an FBI report, he had worked as a bodyguard for a big-time bookmaker with ties to the mob. But following a dispute over money, Billy Veasey pistol-whipped the bookie. Did it, in fact, in front of the guy's girlfriend.

The beating was an insult, a humiliation; and when the bookmaker threatened to use his connections with the Merlino mob to get even with him, Billy Veasey reportedly told him, "Fuck Joey."

That might have been one of the reasons for the hit—but certainly not the only one. Billy Veasey had a history. He also had a brother. Both figured in the shooting.

Three gunmen ran up on Veasey as he pulled to a stop near the corner of Bouvier Street and Oregon Avenue. He never had a chance. Shots ripped through the windows and into his chest, arms, and legs. One tore a hole just below the tattoo over his heart, the one that read I'LL ALWAYS LOVE MY MOMMA.

Billy Veasey was able to gun the engine, but couldn't get

out of the line of fire. The GMC Jimmy lurched across Oregon Avenue and crashed into the front of a neighbor's home. The gunmen disappeared.

This was on the morning of October 5, 1995. Not coincidentally, Veasey's younger brother John was supposed to take the witness stand in the racketeering trial of John Stanfa that morning.

John-John Veasey had been a member of the Stanfa organization. In fact, he was one of the shooters in the hit that left Mike Ciancaglini dead and Joey with a bullet in his ass. But now he was a cooperating government witness. Like his brother, John Veasey was a genuine tough guy. He had the scars to prove it. A year earlier he had been shot in the head after Stanfa found out he was cooperating. With two bullets rattling around inside his skull and with another in his chest, Veasey fought off his two assailants, slashing one across the face with a knife. He then fled the second-floor apartment where the shooting occurred and raced six blocks before collapsing on the steps of an unsuspecting South Philadelphia woman.

The woman called an ambulance, and Veasey was rushed to an emergency room. Two days later, under heavy FBI guard, he walked out of the hospital and into protective custody. Now, on a rainy Thursday morning in October 1995, he was ready to tell a jury all about it. And also about the other hits and misses, the extortions and shakedowns, the bookmaking and loan-sharking he had been involved in as a soldier in the Stanfa crime family.

In a city that has seen a long list of mob informants, John Veasey proved to be one of the most effective ever to take the stand. A street-corner raconteur, he charmed the jury with his colorful yet matter-of-fact description of his life as a Stanfa hit man and extortionist. Even before he opened his mouth, however, his appearance in court gave the Merlino organization an occasion to send a message throughout the underworld.

It was a well-planned attack. The gunmen came at Billy

Veasey from three different angles; at least four shots ripped into his chest, all fired from close range. Seconds later Billy Veasey was slumped dead behind the wheel of his Jimmy, the engine running, the windshield wipers keeping a steady beat as sirens began to wail in the distance.

More than a dozen frantic 911 calls recorded that day paint a picture of the early morning chaos. One of the first, from Billy Veasey's wife, Darlene, came in at 6:45 A.M.

"Please send the police here!" she cried and screamed into the phone. "Oh, my God!"

"What's wrong?" the dispatcher asked.

"Twenty-six twenty-seven Bouvier Street . . . My huband got shot."

"Who shot your husband, ma'am?"

"My husband's Billy Veasey. John's . . . brother," she said, perhaps realizing even in her panic that that said it all.

The dispatcher urged her to calm down and stay on the line while he connected her to a rescue unit.

"Oh, my God! He's on the corner. . . . I can't hold on. He's dying."

A fire department rescue dispatcher then came on the line, and Darlene Veasey frantically told him, "My husband is Billy Veasey, John Veasey's brother. He just went to go get bagels and there was gunshots. Could you please send some-body as soon as possible?"

"Ma'am, I didn't understand a word you said," replied the dispatcher. "What's the address?"

"South Bouvier Street!" Darlene Veasey cried and shouted into the phone. "Twenty-six twenty-seven South Bouvier . . . Please."

Between 6:45:34 A.M. and 6:48:49 A.M. there were four-teen 911 calls made about the shooting on Bouvier Street. On several, people could be heard crying and screaming in the background.

"The whole neighborhood's out," one female caller told an emergency dispatcher.

"There was a bunch of gunshots," said another.

"I heard shots and there's a car that crashed," said a third.

"There's been a shooting here, about five or six shots," said a male caller.

"Did you see who had the gun, sir?" he was asked.

"No. And man, I ain't goin' outside."

At first the speculation was that the hit was a message from the Stanfa organization, an attempt to shut up or intimidate John Veasey. The timing was perfect.

In fact, authorities now believe, the shooting was ordered by Merlino, with Natale's approval. The motive had nothing to do with the Stanfa trial. The fact that the shooting only made John Veasey more determined than ever to testify was an added bonus for those who set the hit in motion.

Billy Veasey was killed, authorities contend, to avenge the murder of Mike Ciancaglini. And also to let it be known throughout the South Philadelphia underworld that anyone who decided to cooperate with authorities would be putting his family, as well as himself, at risk. What's more, the feds believe, Merlino and Natale tapped John Ciancaglini, the oldest of the three Ciancaglini brothers, to take part in the hit.

A brother for a brother: It was all part of a twisted and bloody tangle of family ties and vengeance that stretched back to the first botched hit on Mike Ciancaglini in 1992. That had led Michael to order the hit on his brother Joe. In response, John Veasey gunned down Michael. And, according to the FBI theory, that led in turn to the murder of Billy Veasey by John Ciancaglini.

City of Brotherly Love indeed.

Shortly after the hit, Natale would later tell authorities, he met with Merlino and several other members of the organization who had been involved. At that meeting, Natale said, John Ciancaglini told him, "It was a pleasure and an honor to do it for Michael and for us."

Only one member of the tightly knit Merlino crew refused to take part in the Billy Veasey hit, Natale said. That was

Steve Mazzone. Mazzone felt that John Veasey "was crazy," Natale told authorities, and that no matter how long it took, even if he spent ten years in jail, John Veasey would eventually come home and avenge his brother's death.

As this is being written, John Veasey is approaching his release date after serving a ten-year sentence. He has told several close associates that he is coming home to South Philadelphia, despite the fact that, as a government witness, he could disappear into the Witness Protection Program. Instead, John Veasey has said he intends to "deal" with those who killed his brother.

It would be several years before the FBI had gathered enough evidence to bring any charges in the Billy Veasey hit, and even then the case would be fragile and poorly constructed. To date no one has been convicted of the murder, although authorities are convinced they know all the details.

Previte could offer little help. Despite the fact that he had established ties with Natale, he was still an outsider when it came to Merlino and the crew Skinny Joey kept around him.

At first that didn't seem to matter to the FBI agents running the mob investigation. Natale was the boss, and he was clearly the primary target. In September 1995, they had begun bugging Natale's home phone. Later that fall, by the time Stanfa and his associates were on trial, the FBI had bugs planted in Natale's penthouse apartment and in the Currier & Ives Room at the Garden State Racetrack in Cherry Hill.

Between Previte and the electronic listening devices, the FBI knew virtually every move the mob boss was making.

It was almost too easy.

Shortly after taking over as boss in early 1995, Natale drove up to New York City for a meeting. Big Billy D'Elia, the alleged leader of what was left of Russell Bufalino's old Scranton–Wilkes-Barre crime family, had set it up. D'Elia was going to vouch for and introduce Natale to some leaders of the Colombo crime family. Natale already had connections with the Gambino, Genovese, and Lucchese organizations from his days in prison.

It was a classic gangster excursion: a ride up the New Jersey Turnpike in a fancy car. Drinks at the bar in the Palace, a posh midtown Manhattan hotel. Then a drive out to Brooklyn and a lavish dinner in a sprawling basement-turned-taverna of a double row house owned by a high-ranking member of the Colombo organization.

Twenty to thirty gangsters gathered that night to break bread with the new boss of the Philadelphia mob. Natale, out of prison less than a year, loved it.

So did the FBI.

A group of agents had tailed Natale and D'Elia to Manhattan and then over to Brooklyn. As the wiseguys ate and drank and shared bonhomie, agents outside kept a log of everyone who came and went.

It was all part of an airtight case the FBI was building against Natale, twice-convicted narcotics dealer and self-described Mafia don who would eventually admit involvement in a dozen gangland shootings, eight of which ended in murder. Two of those hits Natale had carried out himself. The others he planned or authorized.

"I never got any happiness or satisfaction out of killing anybody," Natale later told authorities. "I did it because it had to be done."

The most brutal was the 1973 murder of Joey McGreal, a one-time Natale ally in the Bartenders Union. At the time, McGreal was making noises about taking over the labor organization that Natale and Angelo Bruno were then controlling.

While some describe Natale, even to this day, as little more than a gofer for Bruno in those early days, Natale has always insisted that he had a special bond with the legendary don. "He was the boss of La Cosa Nostra. . . . I was his dog," Natale said with typical bravado during one of his first appearances on the witness stand. "When he had a problem and it was serious, he would set me free."

Natale, among other things, may be guilty of plagiarism. The quote sounds suspiciously similar to Sammy the Bull Gravano's description from the witness stand of his relationship with John Gotti: "When he barked, I bit."

On Christmas night, 1973, Natale invited his friend Joey McGreal out for drinks, pulling him away from his wife and two children. Natale lured McGreal to the parking lot of a restaurant; then, from the backseat, he pumped two bullets into his head. McGreal was found slumped over the steering wheel the next morning.

It was something that "had to be done," Natale would boast. That McGreal was the godfather to one of Natale's children was immaterial. The only family that mattered to him, he said, was the crime family, La Cosa Nostra. Even though he was not a made member of the Bruno organization, Natale said he always conducted himself like one.

"Ralph might have bullshitted about a lot of things," Previte says, "but he was a stone cold killer. That was no bullshit."

Out of prison less than a year, Natale was clearly in the sights of the FBI and U.S. Attorney's Office. Conspiracy, extortion, bribery, loan-sharking, gambling, and drug dealing: the feds would get it all on tape. The charges in the new drug case alone would have guaranteed that Natale spend the rest of his life in jail.

But somewhere along the way, the emphasis of the investigation shifted. For reasons that may never be completely understood, it became more important to the FBI to bring down Skinny Joey Merlino than to nail Ralph Natale.

Merlino, young, brash, and arrogant, had come to epitomize everything those running the FBI's Organized Crime Squad in Philadelphia abhorred. Getting Joey—wiping the silly smirk off his face, sticking his fancy clothes and lavish lifestyle up his ass—became as important as building a criminal case. It wasn't about justice. It was about winning.

It was personal. The FBI had a hard-on for Joey Merlino.

And his amazing ability to not only survive but prosper, in the midst of a mob war and a federal investigation, only intensified the animosity.

The Mikey Chang shooting, in which Merlino escaped with a flesh wound, was just one example. Stanfa hit man Phil Colletti later testified about a dozen different occasions when he planted a bomb or laid in wait to ambush the young wiseguy. Something always happened to thwart the plan: the detonator malfunctioned, the cops showed up; Merlino didn't.

All his life, it had been like that for Joey. Before Stanfa, Nicky Scarfo had wanted Merlino dead. Merlino had outlasted them both. He was a survivor.

Now Skinny Joey was positioning himself to tap-dance away from any problems Natale would have. He kept his distance, occasionally sending money but seldom getting directly involved with the boss.

Even the murders that they plotted and approved were orchestrated to keep Natale in the dark. To this day, there are those familiar with the way Merlino operated who insist that Natale was never told the truth about the way any hit was carried out. But Joey Merlino's staying power was about more than beating cases and dodging bullets. As he had told Natale after the Mike Ciancaglini hit: he had the devil in him.

He got the nickname "Skinny Joey" for obvious reasons. He was a small, scrawny little kid. That plus the fact that his cousin, who was also named Joseph, was chubby. They both lived around Tenth and Jackson Streets in South Philadelphia, and so to distinguish between the two there was "Skinny Joey" and "Fat Joey." Later, the cousin moved to the Atlantic City area. The nicknames stuck.

Skinny Joey stayed in South Philadelphia. He went to Epiphany of Our Lord grammar school and then Saint John Neumann High School, the Catholic high school for South

Philly residents. He was bounced out during his senior year for "disciplinary reasons," his mother, Rita, later explained. But not, she was quick to add, "until after they collected the tuition payment." He went on to graduate from South Philadelphia High School.

By that point, though, Joey Merlino was already looking beyond academics. As a teenager he worked as an apprentice jockey, mucking stalls and giving horses their morning workouts at Philadelphia Park, the local racetrack located in the Northeast section of the city. Eventually he even got to ride in races at Philadelphia Park, the Atlantic City Racetrack, and Pimlico.

"He was good," Rita Merlino said, proudly recalling a race he won at Pimlico. But he was also growing. At five foot five, tipping the scales between one hundred thirty and one hundred forty pounds, Skinny Joey had gotten too big to jockey.

Soon, however, other opportunities would present themselves. Merlino turned eighteen a few days before mob boss Angelo Bruno was gunned down in March 1980. Within a year, his father's best friend in the underworld, Little Nicky Scarfo, took over the Philadelphia crime family. Salvatore "Chucky" Merlino eventually became Scarfo's underboss. Salvatore's brother Lawrence, who later became a government informant, rose to the rank of capo. At the time, young Joey was part of what law enforcement would derisively refer to as the "JV Mafia," the junior varsity of the mob. Other members of that group included Mike and Joe Ciancaglini, Tommy Horsehead Scafidi, Nicky Scarfo Jr., and George Borgesi. They were the next generation of the mob. They would run errands, serve as gofers, and hang on the corner outside the mob clubhouse at Camac and Moore Streets in South Philadelphia.

Nicky Scarfo was the boss, but he was based in Atlantic City. Chucky Merlino, Joey's father, quickly became the top mobster in Philadelphia. The only other gangster with as much status and clout was Salvatore Testa, a charismatic young Mafia prince who, like Joey Merlino, had grown up in the organization.

Salvie was a few years older than Joey. His father, Philip "Chicken Man" Testa, was the guy Springsteen wrote about. In the aftermath of the nail bomb explosion that killed the Chicken Man, Scarfo took over the family and set out to avenge the murder of his friend and mentor. Two mobsters were killed and another was banished from the city before Scarfo was satisfied.

Salvie Testa, who took part in both murders, became a capo in the Scarfo organization. Young and handsome, he was surrounded by his own group of loyalists, a situation that did not sit well with Scarfo. But Chucky Merlino was supporting Salvie Testa, who was soon to be family: Salvie was engaged to marry Merlino's oldest daughter, Maria— Joey's older sister.

The dark-haired and beautiful Maria Merlino was probably the most intelligent member of the Merlino family. She would eventually marry a banker and move out of South Philadelphia—breaking away from the wiseguy orbit that destroyed so many who were close to her.

Joey, as a teenager hanging on the edge of the mob, watched it all unfold.

"This kid grew up in Cosa Nostra," said Nick Caramandi, the Scarfo crime family soldier who became one of the first and most devastating mob informants in the city. "He knew all the moves because he was always around it. He knew it better than some guys twenty years older than him."

The voluminous testimony of guys like Caramandi, Phil Leonetti, and Salvatore "Sammy the Bull" Gravano has exposed the once secret and supposedly sacred codes of the mob. The making, or initiation ceremony, where a mobster swears to live and die for "the family," has been a part of so many books and movies that it is almost a cliché. And *omerta,* the time-honored code of silence that for generations was the shield behind which the organization operated, has been shattered so many times in the past ten years that it is no longer relevant. But when Joey Merlino was coming up in the organization, he saw that these things still meant something. He also saw something else in the everyday workings of an organized crime family. He saw that the business of the mob is often very personal.

The wedding of Salvie Testa and Maria Merlino was set for the last Saturday in April 1984. The gowns were already purchased and had been fitted. The banquet room had been reserved. The invitations were about to hit the mail. There would be nearly seven hundred guests including all the top wiseguys from the city—and emissaries from several of the big New York families and from as far away as Boston and Florida.

In Philadelphia it was billed as the Mafia wedding of the decade, perhaps the century. The marriage of a Testa and a Merlino: it seemed almost medieval, the joining of royalty in a love match that solidified a family alliance and ensured stability within the organization for another generation.

"They were gonna have a big affair, at the Bellevue [the posh Bellevue Stratford Hotel in Center City]," Caramandi said. "They were even talking about trying to get Stevie Wonder or somebody like that to perform. I think they coulda done that. They had a lotta connections."

But Salvie Testa, the fearless mobster who had taken a bullet and nearly lost an arm for the Scarfo crime family during the bloody mob war that followed his own father's death, decided at the last minute that he did not want to be married.

Chucky Merlino, the underboss, was livid when Salvie called it off. Scarfo, who already viewed the young Testa as a potential rival, saw an opportunity. The insult to the Merlino family, he said, could not be allowed to go unpunished. In September, nearly six months after the canceled wedding date, Salvatore Testa was shot and killed in a candy store on

Passyunk Avenue in South Philadelphia. He had been lured there by one of his best friends, supposedly to settle a dispute over the collection of a debt. The friend, Joey Pungitore, had been told by Scarfo that unless he set Testa up, he and other members of his family would be killed.

Testa's body was wrapped in a carpet and dumped on the side of a road in a rural area of South Jersey, where it was discovered several days later.

Thereafter, whenever Pungitore showed up at a bar or restaurant in South Philadelphia, some member of the Scarfo organization would walk over to the jukebox and put a quarter in the machine. Then, over the din of the crowd, you'd hear Dionne Warwick's "That's What Friends Are For."

Scarfo, Chucky Merlino, Pungitore, and a dozen others would eventually be convicted of racketeering charges that included the Testa murder. It was one of nine homicides and four attempted murders that were part of a high-profile 1988 racketeering case. Sentenced to fifty-five years in federal prison, Scarfo is currently in a penitentiary in Atlanta. So is Chucky Merlino, who got forty-five years.

During Scarfo's bloody reign as mob boss, about thirty mob members and associates, including a seasoned group of potential underworld leaders like Salvatore Testa, were killed. Two dozen more, like Scarfo himself, were tried and convicted.

Before he went to jail, Scarfo had a falling-out with the Merlinos and threatened to have every member of the family killed. It was the first time, but not the last, that a sitting mob boss targeted Joey Merlino.

Nick Caramandi, after he started cooperating with authorities, provided investigators with all the details of the inter-family squabbling. "I remember we told Joey—he used to hang out at the clubhouse at Camac and Moore—we said, 'Don't come around here no more, kid. There's nuthin' we could do for ya.' Whenever Scarfo got something in his head, you couldn't get it out. He wanted them dead."

Before any of that could happen, however, Scarfo and the others were arrested, indicted, and eventually convicted. Skinny Joey had dodged his first bullet.

"Joey is for Joey," said another Scarfo-era mobster who became a government informant. "Always was, always will be. These kids around him don't get it. Maybe when they're sitting in jail they will."

One of the first to realize was Tommy Scafidi, who saw the treachery and deceit up close. Scafidi had grown up around Skinny Joey but was never a close friend, never part of his circle. In 1989 Merlino wanted Scafidi to set up Nicky Scarfo Jr. At least that's what Scafidi has told authorities. The hit was supposed to make it clear to everyone that Scarfo Sr., the imprisoned mob boss, no longer had any authority. It would also be, on a personal level, a way for Merlino to reply to the threats that Scarfo Sr. had issued before going to jail.

Scafidi said that Mike Ciancaglini and Merlino wanted Scarfo Jr. dead—that they'd tried to recruit him for the hit during a meeting at the Saloon, a popular South Philadelphia restaurant, early in the fall of 1989.

"Joey and Michael are sitting upstairs at a table," Scafidi said. "I sit down with them. I stay for maybe a half hour. I get hit with the 'good guy, bad guy' routine."

Merlino told Scafidi that he and Mike Ciancaglini were "taking over the city." "We're gonna kill Nicky's son," Merlino said. "We want you to help us." Then Mike Ciancaglini, never a big fan of Scafidi's, chimed in. "If you go tell your fuckin' brother [jailed Scarfo soldier Salvatore Scafidi], I'm gonna kill you, your family, and your brother will get killed in jail. . . . And if you go back and warn that faggot Nicky, Australia's not far enough for you to go."

At one time Scafidi had been close friends with Nicky Scarfo Jr., though by the fall of 1989 they weren't that tight. Still, he tried to tell Merlino and Ciancaglini that he couldn't set up the young Scarfo.

Merlino didn't care. He went off on a harangue about Scarfo Sr. Scarfo claimed to be a tough guy, he said—and yet his own nephew and underboss, Phil Leonetti, had "gone bad," becoming a government witness in 1989. (So had Merlino's uncle, Lawrence, but apparently that wasn't a point Scafidi felt comfortable bringing up at the time.) "That motherfucker wanted to kill me, bury my sisters, my father, my uncle," Merlino said of Scarfo's plan to take out the Merlino family. "And now I'm gonna get him. He can't do nuthin' in the city no more."

Scafidi said he finished his drink and left. Merlino said they would be back in touch in a few days, but after the meeting, Scafidi dropped out of sight.

He later testified that he didn't want to be involved because he didn't trust Merlino and Ciancaglini, and because he didn't want to put his own family at risk. Murder, he said, was part of La Cosa Nostra. But murder, in Scafidi's twisted underworld morality, had to be justified. He didn't think it was right for Scarfo Jr. to be targeted because of something Scarfo Sr. had said or done.

"I didn't think Nicky Jr. deserved to get killed," he said. "He might have deserved to get beat up with an aluminum baseball bat [Scafidi and Scarfo Jr. had once beaten up Marty Angelina with bats on Scarfo Sr.'s order] . . . but I didn't think he deserved [to be killed]."

Merlino and Ciancaglini apparently thought otherwise.

Dante & Luigi's is a small, nondescript neighborhood restaurant on South Tenth Street that has served home-style Italian dinners to generations of South Philadelphia residents. Nicky Scarfo Jr., who lived at the shore, would go to the restaurant for clams and spaghetti in white sauce whenever he came into the city. That's what he was having for dinner on October 31, 1989, when a man wearing a Halloween mask and carrying a trick-or-treat bag walked in. The man in the mask walked up to Scarfo's table, pulled a

MAC-10 machine pistol out of his bag, and opened fire. Scarfo was hit eight times in the arms and body. Miraculously, not one shot hit a vital organ.

The gunman fled the restaurant, dropping the gun as he bolted out the door. Scarfo was rushed to a hospital. Two weeks later, he walked out under his own power. He survived, but the shooting had made the point. His father was no longer in charge of the Philadelphia mob.

For the past fifteen years, Joey Merlino has been the primary suspect in that shooting. Several informants, like Scafidi, have linked him to the attempted hit. So have the Scarfos, father and son. In a secretly taped telephone call made while George Fresolone was cooperating with the New Jersey State Police in 1990, authorities heard both Scarfos, senior and junior, express the opinion that Merlino was the man in the mask.

"He's a snake, that kid," Scarfo Sr. told his son.

Later they discussed a plan to take Merlino "to dinner," a clear reference to killing him.

Scafidi has offered another piece of inside information that says as much about the Philadelphia mob as it does about the hit itself. When Nicky Scarfo was boss, Scafidi said, he loved to watch *The Godfather.* Not only did he find it entertaining, he also considered it something of a training film.

Taking his cue from the film, Scarfo insisted that whenever anyone in his organization carried out a hit, he should make a point of dropping the gun at the scene. This, Scafidi explained, was Scarfo's way of paying homage to the first *Godfather.* He was especially taken by the scene in which "Michael Corleone shoots the police officer in the restaurant," Tommy Horsehead said.

Having seen how effectively recovered murder weapons were used as evidence in the racketeering cases that brought down his father and most of his father's contemporaries, though, Joey Merlino "felt it only helped the government" to leave a weapon behind.

But he made an exception in the Dante & Luigi hit, Scafidi said.

"Merlino purposely left the gun behind because he wanted to send a message to Nicky Scarfo Sr.," Scafidi told the FBI.

Despite that information, authorities have never been able to come up with enough evidence to make the case. Mike Ciancaglini, the only other person who may have had first-hand knowledge of the shooting, is dead. Scafidi and others who have talked only know what they were told, or what they have heard.

For a time, the Halloween night hit—carried out in a crowded restaurant in front of dozens of innocent patrons who could have been caught in the line of fire—was considered one of the most notorious and wantonly violent shootings in the history of the Philadelphia underworld.

Now it takes a backseat to the August 31, 1993, attempt on the life of John Stanfa, the drive-by shooting in the midst of rush hour traffic on the Schuylkill Expressway.

Both were the work, the feds believe, of Skinny Joey Merlino. Each epitomized the arrogant, flamboyant, who-gives-a-damn style of the young South Philadelphia celebrity gangster. They were examples, say the feds, of the way Merlino did business.

So was the Louie Turra "situation."

The problem surfaced in late summer 1995 and would be an issue for several years as Merlino and Natale expanded their control in the South Philadelphia underworld.

"Louie Turra and Joe Merlino had a rivalry going for quite a few years," Ralph Natale told a federal jury. "It started over girls, over who's better looking, who's got more clothes. It was a constant rivalry thing between the both of them. The only difference between the both of them, Joey never earned his living."

Louie Turra was the son of Anthony Turra and the nephew

of Rocco Turra, two brothers who knew their way around the underworld. Independent, fearless, and entrepreneurial, they had operated on the fringes of the Philadelphia mob for years.

Rocco, the younger brother, had been arrested more than thirty times, with eleven convictions. Among other things, he was a suspect in a notorious mob-linked murder of a Teamster official in the early 1970s. He was also something of a South Philadelphia legend. He once got into a street-corner brawl with four brothers at the same time. He had beaten three when the fourth pulled a gun, shot him, and then split his head open with a hatchet. Somehow Rocco survived, and he wore his scars like badges of honor.

During the Scarfo era, the Turras had balked at paying a street tax, and young Louie had nearly been killed. Only Rocco's intervention—his reputation gave him standing—had saved his nephew. Now Joey Merlino wanted a piece of the action—specifically, of the highly lucrative cocaine operation that young Louie Turra and a group of his associates were running.

Again, the Turras balked at paying.

"Who the fuck does he think he is?" was the common expression heard on both sides of the dispute.

Louie Turra was unimpressed by Merlino's newfound status as mob underboss. Surrounded by his own organization of young toughs, he had no intention of submitting to a street tax.

He was, however, interested in making money. And when John Ciancaglini approached him with a loan-sharking offer, he jumped at the chance. Ciancaglini asked to borrow twenty thousand dollars, money he said he was going to put on the street. Turra lent the money for a point—one percent interest per week. Ciancaglini, in turn, said he was going to put the money out at three points. There was profit in it for everyone.

Instead, Ciancaglini and Merlino put the money in their

pockets. It was Joey's way of shaking Turra down. And it just added to the growing animosity between the two hot-headed young mobsters.

The two exchanged insults through third parties for several months. Then, one night in May, Turra had the misfortune of walking alone into an after-hours club where Merlino and several associates were partying.

The place was called Libations. It was a bad place for Turra to be.

There are conflicting accounts of what happened, but the end result was Louie Turra lying beaten and bloody in the gutter outside the club, stomped—investigators believe—by several Merlino associates as Joey looked on.

At least one account included this data: to add insult to injury, the assailants reportedly took Turra's Rolex watch and "sold" it to a bartender at the club. The price for the piece, worth somewhere from fifteen to twenty thousand dollars, was twenty-five bucks.

It was both another insult and another humiliation.

"It's one thing to give a guy a beating and knock him unconscious," Natale said. "They took his watch off of him while he was unconscious. . . . He had a beautiful Rolex. They took it right off him. . . . He was more angry about that when I finally talked to him."

Turra, naturally, wanted revenge. He began to plot Merlino's demise—unaware that the FBI was getting it all down on tape.

Turra's drug ring was already the target of a federal probe. The feds had wiretaps in place on several phones, and two major informants working from inside the Turra organization. Each move Turra planned against Joey ended up in an FBI file.

The plots, a federal prosecutor would say later, involved "a dangerous mix of firearms, narcotics, and attempted murder."

The first associate Turra tapped to kill Merino turned out to be one of the FBI's informants. Turra and one of his top crew members, Gaetan Polidoro, delivered a Sig-Sauer 9-millimeter semiautomatic loaded with hollow-point bullets to the would-be hit man, who immediately tipped the feds.

Over the next three months, half a dozen plots to kill Merlino would be launched by members of the Turra crew. In each case, though, the feds and the Philadelphia police managed to short-circuit the action. Without realizing it, Joey Merlino owed his life to the FBI.

On June 8, 1995, for example, agents on a wiretap overheard Turra summoning Polidoro, telling him to "come prepared." Acting on information provided by the feds, the police intercepted Polidoro as he was driving to Turra's. They pulled him over in his red Chevrolet Blazer and searched the vehicle: inside they found a MAC-10 automatic .45-caliber machine pistol with a large capacity magazine loaded with jacketed hollow-point bullets.

Two days later, Frankie Russo, another member of the Turra organization, was shot to death near his South Philadelphia home. Turra, authorities now say, mistakenly believed Russo was the "snitch" who provided the information that led to Polidoro's arrest. In fact, the FBI's informants were still very much in place and providing detailed accounts of the plots to kill Skinny Joey.

Later that month, one of the informants accompanied Turra and Polidoro to Miami, where Turra met with one of his Colombian cocaine suppliers. They talked of drugs and weapons. Turra was clearly gearing up for a full-scale war with the Merlino organization.

Polidoro, who was Turra's cousin, was a volatile young drug dealer and regular user of the product he sold. He made up a list of weapons they were looking for. Turra reviewed it and then handed it over to the Colombian. The list was seized a month later when the DEA raided the Colombian's apartment.

The weapons Turra was seeking included six .45-caliber MAC-10 machine pistols, ten 9-millimeter Barettas, and ten

.45-caliber Sig-Sauers. Not one to let his problems get in the way of his business, Turra also asked the Colombian to increase the amount of cocaine he was sending. Turra said he wanted about fifty kilograms a month, roughly one million dollars in coke. "Stepped on" [diluted] and packaged for small street-level sales, fifty kilograms would have generated several million dollars for the Turra group each month.

Louie Turra had no intention of sharing any of that cash with Skinny Joey.

The plotting continued throughout that summer. In late July, Turra gunmen planned to pop Merlino at the Rock Lobster, a hip nightclub along Delaware Avenue where Merlino and his associates often partied. An informant described how the shooters completed a "dry run" at the nightclub, then headed back when they knew Merlino and his entourage had arrived. But the hit men never got a clear shot at Joey and called the attempt off.

A month later, some older members of the Turra group interceded with Natale and the hostilities ceased. "Tony Turra came to see me at the racetrack," said Natale, who had grown up around the corner from the Turra brothers in South Philadelphia. "We knew each other since we were boys. And he told me this whole story. He said, 'Ralph, I want you to save my son's life. He's involved with a lot of people here. Somebody is going to get killed and it might be my son. Maybe we can sit down and talk this over. We only trust you.' "

Natale, playing the Godfather role he so enjoyed, called a meeting at a restaurant in Deptford, New Jersey. Natale brought Stevie Mazzone. Anthony Turra showed up with his son Louie and with Dennis Virelli, a veteran drug dealer who had sided with the Turra group in the dispute.

But there the Don Corleone image ended. For one thing, the gathering spot was the Ground Round, a glorified hamburger joint, hardly worthy of the Don Corleone performance Natale would describe.

"We sat down," Natale said. "I made one big statement. I said, 'Listen, no use telling who's right, who's wrong. Everybody's wrong at this position. . . . I came here to settle this. I grew up with Tony, but I'm with Joey now. . . . I don't want to see nobody get hurt. I don't want to see Louie Turra get hurt; I don't want to see Joey get hurt. . . . I'll stand for the peace here, because if anybody breaks it, then they got to contend with me and everybody I got with me. We have to understand that. We shake hands here now. We part as friends.' And that's exactly what we did."

Previte heard about the "peace" negotiations shortly thereafter and reported back to the FBI. "It was agreed that whatever happened had happened, but that it was over," he said. "Turra didn't have to pay no money, and nobody was gonna try to shoot Joey."

Louie Turra and his organization, however, remained the target of an ongoing federal probe. He, his father, his uncle, Polidoro, and several other members of his crew would eventually be indicted, and later pay a severe price for plotting against Merlino.

Joey, meanwhile, continued to move around South Philadelphia like a man without a care in the world.

He had a beautiful girlfriend, Deborah, a tall, thin Korean-American woman with runway-model looks. And he had a score of groupies, young girls from all over the city, anxious to hook up with him.

"The only things he cared about were bettin' sports, chasing broads, and busting balls," said a friend from that period. "I think after Mikey Chang got killed, Joey didn't care about anything else. He loved Michael like a brother. After Michael died, I think Joey said, 'Fuck it.' You know, he coulda been killed that day just as easy. So it was like he was playing with the house's money. Whatever happened, happened."

Merlino had an entourage of loyalists, guys he had grown up with, who helped him form an organization within the organization. George Borgesi, Marty Angelina, Steve Mazzone—no one from that inner circle has ever cooperated with authorities. It's not about the mob and *omerta*. It's about being from the same South Philadelphia street corner. You don't rat on your friends.

But Joey also had something else going for him: timing. Incredible timing.

Take the day Stanfa was convicted, for example. The trial had lasted about three months. The jury was out for more than a week. No one could have predicted exactly when the verdict would be announced.

But the day the jury did return with its guilty verdicts—November 21, 1995—was the same day Skinny Joey had picked to host his first Christmas party for homeless children. It could have been an episode from *The Sopranos,* though it took place four years before that show hit the air. That was the way Joey Merlino lived: you couldn't script it any better.

Merlino arranged to rent a restaurant on Passyunk Avenue, just down the street from his Avenue Café. He raised money, he said, from merchants up and down the avenue, and from other friends and businessmen who wanted to help out.

That November afternoon the restaurant put out a spread, starting with escarole soup (an Italian holiday tradition), then rigatoni, then turkey with all the trimmings. There was a huge decorated Christmas tree in the corner. Piles of presents wrapped in bright holiday paper were stacked under and around it—dolls, racing car sets, boom boxes, tape recorders, basketball and football jerseys.

Merlino had arranged for forty-five residents of a local homeless shelter to be bused in for the lavish party. When the bus pulled up outside the restaurant on Passyunk Avenue, jovial mob associate Angelo Lutz was there in a Santa Claus costume, waving and cheering the kids on. Behind the five-foot-four, four-hundred-pound Lutz was a South

Philadelphia string band in full regalia serenading the kids with a banjo-strumming, saxophone-wailing version of "Santa Claus Is Comin' to Town."

Merlino, dressed in a glen plaid suit over a collarless white shirt, smiled from the sidelines as he watched the festivities begin. The media, tipped in advance, were there in full force.

"It's for the kids," Joey said in a brief statement before turning the affair over to his lawyer, who handed the head of a homeless rights group a donation of thirty-five hundred dollars.

"We just want to thank the Merlino family," she said with a straight face as Merlino's associates cheered and reporters rolled their eyes. That night on the six o'clock news broadcasts, and the next day on the front pages of the city's papers, two stories vied for top billing. "Stanfa Convicted," screamed one headline. "Merlino Feeds the Homeless," shouted the other.

The media were creating a monster, said some in law enforcement. Merlino, a crook, a gangster, a suspected murderer, was becoming a celebrity.

But at that point—on paper at least—Ralph Natale was still the primary target of the FBI.

6

It was seamless, astounding. It was the FBI at its best.

One boss had just been convicted, largely on the evidence of his own words, recorded in conversations he never assumed anyone would overhear. And even as that case was rushing to its dramatic finish, the feds were already up and running with another electronic bugging operation that had the new boss literally wired for sound.

Almost anything Ralph Natale said or did over the next three years became part of an FBI report, gleaned from one of two sources: the FBI's own listening devices or the inside information provided by Previte.

The FBI placed its first bug on Natale's phone in July 1995. Bugs were planted in his home and at the racetrack in November, just as a federal jury was deliberating Stanfa's fate. The recordings continued through October 1996.

On those tapes, the feds heard Ralph talk about dozens of deals and scams he was planning to put into place. They ran the gamut from rigging bids for government contracts—Natale talked at length about how he was bribing the mayor and a councilman in the city of Camden—to loan-sharking, extortion, and gambling. They included quasi-legitimate enterprises in which mob-backed companies would sell cigars, food, and produce to the Atlantic City casinos, import tomatoes from Canada, collect medical debts for doctors' offices, and operate an Internet pornography site.

The talk went on for weeks at a time. Natale always had

three or four balls in the air, was always about to close a deal. None of it ever happened.

None of it.

The only real money Natale was able to generate was through the one business he truly knew and understood— drugs. Strip away all the talk, all the posturing, all the bravado, and that's what was left: Ralph Natale, narcotics trafficker.

But the game had to be played out.

And so here was Natale in May 1996, explaining to an associate on the phone about the Internet porno company, modeled after a business in California that had twenty-five thousand members. For a fee, a member dialed up the service; then, from the privacy of his own home or workplace computer screen, he could "look at broads." Broads with guys, broads with other broads, broads with animals or toy devices. Broads in pairs, in trios, in groups.

In another conversation he discussed opening a strip club at the New Jersey shore. In several others he talked about massage parlors and whorehouses.

Previte, ever the opportunist, picked up on the talk.

"I had met this broad Sherry," he said. "She was half Filipino and half cheese-head, you know, from Wisconsin. Real good-looking. And she was talking about going to work for an escort service. So I suggest to Ralph that maybe we could go into business with her. And then I tell the FBI that Ralph wants to open a whorehouse."

Previte and Sherry scouted locations, finally settling on an apartment in Center City, a few blocks from the federal courthouse and City Hall.

"She and I set it up," Previte explained. "She got some of the girls she knew to come work for us. We got this apartment in Philadelphia, at Eleventh and Waverly. I used a stolen credit card to buy furniture for the place. We went to the furniture store that Danny Daidone [a Natale associate] had up in Northeast Philly. We called the place The Green

Door, 'cause the building had a green door. We had ads in some of the weekly papers. I think we could have made some money. . . . Freddy [Aldrich] was running the place for me. For a while we did all right. There were some lawyers who used to come in regularly. If I mentioned the names, you'd know them."

The business was run out of a first-floor apartment in a well-kept two-story brick residence on a quiet side street. The ads that Previte placed in the weeklies told the story. "The Green Door. New Girls! Young Girls! College Girls! Guaranteed Good Time. Private CC [Center City] Location. Or Yours. Seven Days a Week." The ad included a local phone number.

"A customer would call and we'd direct him to the location," Previte said. "Or we could send a girl out to him."

Previte also had business cards made up: plain white cards with green lettering listing The Green Door and two phone numbers.

"We coulda made some money," he says.

But there were problems with the business almost from the start. "Some of the guys used to come in and screw the girls for free," Previte said. "I still hadda pay 'em. And every time Ralph had a party he wanted me to send some broads over. 'Course he would never pay for them either."

What could have been a moneymaker—as it's always been—instead turned into a short-lived opportunity for Natale and some of his associates to score some free sex.

"I hadda shut it down," Previte said. "I was losing money, and it was just a headache. I didn't need it."

Natale's plans to generate millions of dollars in construction contracts through the city of Camden also went awry, despite months of planning and dozens of secret meetings with city officials.

When Natale met with a city councilman at The Pub, a popular South Jersey restaurant, and later arranged for the councilman's girlfriend to get a four-thousand-dollar diamond ring as a gift, the FBI was there watching and listening. When he ordered his "chief of staff," Danny Daidone, to wine, dine, and bribe soon-to-be Camden mayor Milton Milan, the FBI's tapes were rolling.

Cash in envelopes, and vacation trips bought and paid for by the mob, were part of Milan's end of the deal. But the construction companies, which Ralph hoped to use to suck money out of the city, got squat.

The deals never came through.

The planning, however, did offer a rare and candid explanation of how the mob preferred to benefit from construction deals. Natale, ever the expert, described the technique to an associate as "shaving." "You cut down to the bone," he said. "Everything has to be inferior materials. When I need four wires, I put three. Twenty nails, I put ten."

Later, in another conversation, he told the same associate, "We're not building the pyramids. If it only lasts five years, that's not a problem. . . . Cosmetically, it'll look good."

Looking good was always important to Ralph Natale. But the fact is, during his three years as Philadelphia mob boss, the only thing he built was a drug operation.

He talked about a lot more, of course. Over and over and over. On the phone, and during meetings with associates at his home and at the racetrack, Ralph Natale went on at length about everything he was going to do.

"This was Ralph being Ralph," a defense attorney would later jibe. "He was a legend in his own mind."

Hard work, Natale liked to say, was the secret to success. And the kids, he would complain—meaning Merlino and his associates—just didn't get it. "Nobody wants to go to work," he said. "They want to be gangsters. And they can't even be that."

In another conversation, a Natale associate complained about Merlino and about the media attention he was receiving. The associate said he had overheard Merlino and some of his crew talking openly about "Mafia, Mafioso, and Ralph this and Ralph that.

"It's not right," he said. Natale nodded in agreement.

The associate, however, was not without his own vanity. In the same conversation he detailed a meeting he'd had with a top casino executive in Atlantic City, trying to set up one of the many deals that Natale was floating at the time. During the meeting, the executive told the associate that he looked like the actor Joe Pesce.

This didn't sit well with Natale's guy.

"I told him, 'No, Joe Pesce looks like me. I was Joe Pesce a long time before that fuckin' guy. When I was a kid, I was a tough motherfucker. But no more. I'm a businessman now.'"

No one knows how the casino executive reacted to that clarification; suffice it to say that the deal in question never went through. Then again, neither did any of the other moves Natale hoped to make at the shore.

"If we don't become successful in Atlantic City . . . we ought to put weights around our necks and jump in the river," Natale said on one tape while boldly bragging about the fortune he intended to make in and around the casino industry.

At other times, Natale would complain that he wasn't getting enough money from the gambling and bookmaking operations, from the loan-shark business or the extortions. He knew Merlino was holding back, but couldn't quite bring himself to say it. Instead, he lashed out in general frustration.

"Nobody knows how to count," he complained about one gambling operation. "I'm gonna start carrying a fuckin' bat around." Everyone comes to him with complaints, he moaned at another point, but "nobody brings me a little happiness."

Happiness to Natale was an envelope full of cash.

"I got to get that ball-peen hammer," he said. "The Italian hammer. Then when they talk, you give them the circle right in the forehead."

Natale was always going to meetings, always setting up lunch or dinner dates with business associates or investors who might front money for one of his "deals." He wanted to get back in the union business in Atlantic City, and talked up a pie-in-the-sky plan to launch an independent union for the workers at the lavish hotel Steve Wynn was planning to build at the time. Natale also tried to get one of his daughters a job with a Wynn associate, another move that came to naught.

Natale saw the arrangement as a way to get his foot in the door with the Wynn organization, and apparently he had no problem with his daughter going on several dates with the Wynn executive—a man close to his own age—if that would help him get where he wanted.

Natale liked to portray himself as an old-school gentleman, but using his beautiful young daughter to gain access to a casino deal suggested otherwise. So did another secretly recorded conversation in which he flew into a rage about a secretary who put him on hold when he called her boss's office. He threatened to set the woman on fire if she ever did it again.

"Don't think I won't," he said. "I'll come in with a fuckin' gallon of gasoline. I'll pour it all over her."

For months Natale was a "hidden investor" in a bar-restaurant he hoped to open in Cherry Hill. It was the kind of spot he always wanted, a 1960s "rat pack" nightclub with a menu of steaks, chops, and pasta dishes and a well-stocked, darkly lit, and smoky bar. He'd hold court there, just like in the movies. He wanted to call it Pal Joey's, a play on an old Sinatra film, and also on Natale's friendship with Skinny Joey Merlino. It was Natale's way of tweaking the authorities.

Those authorities, however, had the last word. An application to have the property's liquor license transferred to the ownership group that was fronting for Natale crumbled when New Jersey investigators tipped the Cherry Hill police that Natale would be the true owner of the joint. Pal Joey's died on the vine.

Eventually Natale realized that he was under constant surveillance and that his phone and house were bugged. He even got a tip about the bug at the racetrack. He would point to the ceiling, waving his index finger to indicate that the tapes might be rolling.

By that point, though, it was too late. The cases were already being made.

"Everywhere I go, I'm being watched by Uncle Sam," he said in a telephone call a few months before the bugging operation ended. "I can't even talk on this thing anymore."

The FBI was also listening and taping when Natale, Daidone, and another associate took the phone off the hook and tried to determine whether it was bugged. The transcript reads like something out of a Marx Brothers movie. The audio includes scraping sounds, sighs, and occasional banging.

"Turn it the other way. . . . Go back, go back. . . . Okay, you're dead. That's good." All the while, the bug planted in Natale's living room kept running, picking it all up.

In 1996, shortly after it was disclosed that two more Stanfa associates, Sergio Battaglia and Herb Keller, had decided to cooperate, Natale was heard on tape describing them as "pieces of shit." And after a news show featuring former-mobster-turned-informant Phil Leonetti aired on national television, Natale complained to his grown son Michael that Leonetti should keep his mouth shut "and do his time like a man."

All these conversations would prove ironic after Natale cut his own deal with the government. Natale was also captured going on at length about informers ("rats," he called them)—about how despicable they were, how they didn't deserve to be called men.

Natale always boasted that he had done more than fifteen years in federal prison; that he refused the feds' every attempt to get him to cooperate; that as a result he had "maxed out," serving his maximum sentence because he refused to sing. He was a man's man, a true Mafioso, a man of honor.

But Natale was realistic enough to make one other observation about cooperating witnesses.

"I hate them, you hate them," he told an associate. "Everybody in the world [hates them]. Because what they did to themselves shouldn't be done to a man. But very seldom does an informant lie." As he recognized, the risk of losing the immunity deal is too great.

"They have no need to lie. . . . They tell the truth . . . so they can keep their deal . . . and put another man in jail."

One of the things Ralph Natale asked Previte during their first meeting at the Fireside Restaurant was whether he could move any drugs. Neither Previte nor the FBI was surprised by the request. The fact that Ray Rubeo was sitting at the bar when Previte walked in that night was indication enough that Ralph was getting back into the drug business.

Rubeo and Johnny Santilli, a South Philadelphia underworld figure who had begun showing up in Natale's circle, were both well-known methamphetamine distributors. Santilli had been doing it for years: he had been caught up in a big mob-connected federal drug case during the Scarfo years, and had even served a prison stint. Now, like Natale, he was back doing what he knew best.

Rubeo was more circumspect, but he also had a history in the business. Based in Manayunk, a Soholike section of Philadelphia, he ran a legitimate business—a furniture store and warehouse. His company manufactured outdoor furniture, the kind of stuff that ended up on patios and decks all over the Delaware Valley. But his big money was in meth. Rubeo became the source for the first deal Natale set up for Previte.

"Ralph wanted to know if I could move some stuff," Previte said. "I told him sure. It didn't matter to me. The FBI was gonna front the money. It was no problem. . . . I agree to the deal. Originally, it was twenty-five grand. And then the feds say they're not gonna give me the money. They tell me to call it off. 'Fuck that,' I said. 'What, are you nuts?' I gotta do the deal. You can't go halfway with this stuff. That can get you killed. Finally, they agreed. I think they had to go to the DEA to get the funds. But we went ahead with it."

The transaction took place in the parking lot of the Diamond Diner in Cherry Hill. The meeting had been set up in advance. The FBI had surveillance in place. Previte arrived first and was waiting in his car, a late model Cadillac, when Rubeo pulled up in a white Lexus. After Rubeo got out of his car and entered the diner, Previte followed him in. It was a little before 3 P.M.

They sat, drank coffee, and had a brief conversation. Rubeo wanted to know if everything was "all right." Previte smiled.

"Everything's fine," he said.

At 3:25 P.M. the FBI watched as Rubeo and Previte walked out of the diner together. Each headed for his car. Rubeo opened the trunk of his and pulled out a light-colored seat cushion. Previte then pulled up in his Cadillac and popped the trunk. Rubeo put the cushion in the trunk of the Cadillac and slammed it shut. Then Rubeo got in his car and headed back toward Philadelphia.

Previte drove east on Route 70 for about two miles to the parking lot of another diner, where he met up with a team of FBI agents. They then drove both cars to a third location, the parking lot of Champs, a sports bar. There Previte popped the trunk of his Cadillac once more.

In the stilted language of investigative memos, Special Agent James T. Maher, head of the FBI's Organized Crime Unit, described what happened: "I removed a seat cushion of the kind used on outdoor furniture," Maher wrote, "and took possession of the cushion and its contents. I unzipped the zipper and noted that there was concealed inside a block of white material which had been broken into pieces and had been placed inside a Ziploc-type plastic bag within another bag of the same type."

Maher took the cushion and the "white material" back to

the FBI offices in Philadelphia. When the contents were tested, the material was confirmed to be a half-pound of meth.

Previte had paid seven thousand dollars for the eight ounces; depending on how it was packaged and whether it was diluted, it could have generated two or three times that much in street sales. That was the economics of the drug business. Of the seven grand, Rubeo was to get five, and Natale and Anthony Viesti were in line for one thousand each.

It was October 7, 1995. Natale had been out of prison for about a year, and already he was the target of a drug investigation that could send him back to jail for the rest of his life.

One of the great myths surrounding the American Mafia is the organization's supposed ban on dealing drugs. There are some who believe this was a figment of Mario Puzo's imagination, a literary attempt to enhance and ennoble the character of Don Corleone. In truth, the mob has always been involved in the drug business, to one extent or another. Any ban or prohibition instituted by the occasional mob boss was usually based on pragmatism, not morality.

Many old-time mobsters simply believed drugs brought too many problems. For one thing, you were forced to deal with lowlifes. Dealers often became users, and users couldn't be trusted. What's more, narcotics was a socially unacceptable business. It was one thing to make book or take numbers. Everybody placed a bet now and then; where was the harm? Hell, in some places the states themselves sanctioned lotteries, casinos, and horse racing. Even loansharking could be described as a victimless crime. People with bad credit often need to borrow money. If a bank wouldn't provide it, then the mob would.

Of course, there were some wiseguys who believed that drugs were a dirty business, impossible to justify. But there were just as many others who either had no moral compunction about dealing, or who just couldn't resist the cash.

The murder of Paul Castellano by John Gotti, for one, was triggered by drugs. Investigative tapes that captured several top Gotti associates in the middle of highly lucrative narcotics deals were about to be released. Like Angelo Bruno in Philadelphia, Castellano had more money than God, and enough legitimate and quasi-legitimate sources of income to scorn the drug trade and prohibit those in his organization from dealing. Castellano was shot because Gotti knew that if he didn't strike first, he and his associates would have been killed for getting involved in trafficking.

In Philadelphia, too, there were mobsters willing to ignore Bruno's ban on dealing. The ban outlasted Bruno himself, but there were ways around it. During the Scarfo era, the mob extracted a "street tax" from drug dealers. Wiseguys, including Scarfo himself, also lent money to dealers at exorbitant interest rates, in effect financing the drug trade at a level once removed from the streets.

Meth has always been big in the Philadelphia area. At one point in the late 1970s, the City of Brotherly Love was considered the meth capital of the country. The mob, freelance dealers, and outlaw motorcyle gangs were all involved in the trade. Meth labs popped up in old garages in industrial neighborhoods and on farms and in the backwoods in areas of New Jersey and Pennsylvania a few hours from South Philadelphia.

The key to making meth is phenyl-2-propanone (P2P), a liquid chemical banned in the United States but sold in several European countries. Germany was a key source for "oil," as it was called, which was shipped in five-gallon drums.

The now-defunct Pennsylvania Crime Commission, in a 1990 report on organized crime, detailed the economics of the meth business in a section concerning the mob and the drug trade.

"Vast profits," the commission noted, were possible in the methamphetamine business. "The price of a gallon of P2P in

France or West Germany in the early 1980s was $135 to $155," the report noted. "Once the drums reach Philadelphia, a gallon is typically sold for between $2,500 to $7,500 at the wholesale level, a mark-up of between 1,800 and 4,800 percent. . . . A gallon of P2P translates into ten to twelve pounds of almost pure methamphetamine. Over the decade, the price of finished methamphetamine rose from about $6,000 per pound to $10,000 per pound. A gallon of P2P, then, could translate into gross profits of $60,000 to $120,000."

Scarfo was indicted in the late 1980s for financing a deal to import fifty gallons of P2P. His end was going to be a half-million dollars. It was one of the few cases he beat. While charging Scarfo with being a drug kingpin, the indictment ignored the more obvious evidence that Scarfo was financing and extorting drug dealers. There was no evidence that Scarfo or anyone in his organization sold meth, and little to support the kingpin allegation. John Santilli was one of twenty other mobsters and mob associates indicted in that case. He was also one of a dozen who were convicted.

Skinny Joey Merlino, taking a page out of the Bruno-Scarfo narcotics handbook, also kept himself once removed from the trade. Merlino liked the money drugs could provide, but he knew that narcotics trafficking was, in his own words, a "bad pinch." So while Ralph Natale was setting up meth deals, Skinny Joey was sticking to gambling, loansharking, and extortion. But several of the people around him were dealing. And that didn't seem to bother him at all.

Nicholas Volpe, a South Philadelphia entrepreneur who was eventually nabbed for selling cocaine, marijuana, and an assault rifle to an undercover DEA agent, was for a time a regular at the Avenue Café and part of Merlino's social set. Volpe, a city kid, grew his own grass. During a raid at an apartment Volpe owned on Carpenter Street, agents found 182 marijuana plants under cultivation, flourishing in an elaborate hydroponic system that Volpe had installed. Employing timers that controlled the lights, the temperature, and the daily watering routine, the setup was virtually maintenance-free.

Merlino claimed to know nothing about the drugs, but he would occasionally "borrow" money from Volpe, money they both knew Joey never intended to pay back. It was the price Volpe, a thirty-three-year-old wannabe wiseguy, was willing to pay to move in Merlino's circle. There was status in being able to say, "I'm with Joey." Volpe even took it one step further, having Merlino stand as the godfather for his daughter.

Roger Vella, another Merlino sycophant, was also heavily involved in the drug trade. Vella, a high school dropout, was several years younger than Merlino, and idolized the rising mob star. He kept a scrapbook of newspaper and magazine clippings about Joey, whom he referred to affectionately as "Baba." He had pictures of Merlino on the walls of his bedroom. For a time Vella served as Merlino's driver, ferrying him around town in his own car.

Vella, at twenty-five, had no job and no legitimate source of income. He lived at home with his mother, father, and younger sister. Yet he drove a Cadillac SST, and always had a wad of cash in his pocket, frequently picking up the tab while partying with Merlino and his associates.

Like Volpe, Vella was buying his way into Joey's world.

Merlino was happy to take their cash. The fact that they were drug dealers was immaterial to him. And if their association·with him offered them a degree of protection, and provided them with access to even bigger deals and more cash, so much the better. In the end, it was all just more money for Joey. He couldn't have cared less where the cash came from.

There were, of course, residual problems—the kind that old-timers like Bruno and Castellano knew came with the drug business. In February 1995, police found a body hogtied and set on fire along an abandoned street near the South Philadelphia Food Distribution Center.

The victim had been shot several times in the head and back. He was stripped down to his underwear, and a five-dollar bill and two ones had been stuffed into his shorts. Veteran organized crime investigators saw it as a lowlife version of the Antonio Caponigro execution, an attempt to mimic the gesture of leaving Tony Bananas with twenty-dollar bills stuck up his ass.

The victim, in this case, was Ralph Mazzuca. Mazzuca, who owned an auto detailing shop in South Philadelphia, was also suspected of dealing drugs, primarily cocaine. He was a rival of Roger Vella's in the drug trade, and Vella quickly became a suspect in the hit.

What's more, the murder appeared to be tied to the Merlino mob. George Borgesi, a top Merlino associate whose family lived just up the street from Vella, was brought in for questioning on the day the body was found. Police suspected that Mazzuca had been killed in Vella's home, that his body had been hauled from there and dumped near the Food Distribution Center. Investigators hoped to tie the hit to Borgesi and Merlino.

Questioned for several hours and then released, Borgesi insisted he had nothing to do with the murder. What's more, he said he would never be involved with drugs. But over time, in private conversations, he would also argue that if Vella had murdered Mazzuca it was justified.

Sometimes, Borgesi said, street justice is the only justice that matters.

As detectives looked into the Mazzuca murder, they learned that several months earlier Vella's home had been ransacked and his girlfriend had been briefly kidnapped. Two Hispanic men had broken into Vella's house on Tree Street in South Philadelphia, pistol-whipped, tied up, and terrorized his parents and younger sister, and stolen two kilograms of cocaine. Vella wasn't in the house at the time.

A few days later, the same two men were believed to have grabbed Vella's girlfriend, throwing her into the trunk of a car and driving to another part of the city, where she was briefly held for ransom before being released, frightened but unharmed.

Vella never paid any money, and later bragged to an associate that the intruders had taken "two bricks" of cocaine but had missed the cash, estimated at a hundred thousand dollars, that he had stashed in the house.

Both incidents, the home invasion and the kidnapping, stemmed from a dispute between Vella and Mazzuca over drugs and money, detectives now believe. Vella thought Mazzuca was behind the attacks on his family and girlfriend.

"If somebody did that to your family, what would you do?" Borgesi asked more than once.

Borgesi remains a suspect in the Mazzuca murder. Vella has since been charged and has pleaded guilty. He is now cooperating with authorities, underscoring the concerns mob bosses like Bruno and Castellano had about getting involved with narcotics and those who traffic in drugs.

Joey Merlino never inquired about the source of the money people like Vella and Volpe gave him. It was his own don't-ask-don't-tell policy.

1

Ralph Natale, on the other hand, always knew where the money was coming from. He was constantly setting up deals, anticipating his cut. Not only were he, Previte, Viesti, Rubeo, and Santilli dealing, but Natale also got his son-in-law, Robert Constantine, involved.

Natale and Santilli had done time together in the federal prison in Danbury, Connecticut. As fellow convicted drug traffickers, they naturally discussed the business. Santilli specialized in meth. He was a "cooker," someone who knew how to take P2P and turn it into speed.

Santilli, who was released from jail before Natale, got together with Natale's son-in-law after he returned to South Philadelphia. "They put their heads together," Natale said, and worked out a business relationship. Eventually they contacted Rubeo, who also became a part of the operation.

When Natale got out of jail late in 1994, he started in right away discussing ways they could expand their drug activities. It was, Natale knew, the quickest and surest way to make money.

At the time, Natale said he "didn't have a dime." The crime family he and Merlino were trying to reorganize wasn't in great shape either. So in the short term, he "borrowed" fifty to sixty thousand dollars from several business associates, who knew he had no intention of repaying them. Like Merlino's sycophants, they were willing to put out the money in exchange for a chance to be "with" the new mob boss. Natale and Merlino also set about establishing control of the street-level gambling and loan-sharking in the city. Anyone involved in the business was told that he had to make a tribute payment to the organization.

The so-called street tax, first instituted during Scarfo's reign as boss, was the cost of doing business. Bookmakers, loan sharks, number writers, even drug dealers who were known to Merlino and his associates—all were strong-armed. *Pay or get out of business:* that was the message.

"Nobody was freelancing no more," Natale said. "Everything illegal in that city belonged to us."

In addition to a weekly or monthly stipend, everyone was required to deliver a "Christmas package," usually in the form of an envelope stuffed with cash. Depending on the size of the donor's illegal business, the amount could range from several hundred to tens of thousands of dollars.

But it would take months to reinstitute the street tax, to develop information on who should be approached and get a cash flow going. Natale also wanted the organization to establish its own gambling and money-loaning operations. It was the drug trade, he knew, that could generate the most money in the shortest amount of time.

"I wanted to give us a jump start," he said several years later, explaining why he immediately got back into the drug business after his release from prison. "We needed money to back the lottery, the numbers, the bookmaking. We needed a lot of money for loan-sharking. The only jump start I could have got was to get started in the meth business. And we had an opportunity and I saw it and I went forward.

"I managed the entire operation. I made sure . . . what the prices would be, who they would sell to. I controlled it. . . . Everything that was illegal in the city of Philadelphia, South Jersey, Atlantic City. We were going to do what we had to do."

Natale's cookers were Rubeo, Santilli, and his son-in-law. Merlino knew about the drug operation, Natale said, but did not play an active role in it. "I said, 'We're gonna get off. We're gonna be in good shape. We're gonna get enough money.' That's what it's all about, money. I said, 'We'll be fine.'"

Natale promised Merlino a piece of the action once the drugs and money began to flow. Merlino was always looking for money.

"He was a great gambler," Natale said, then corrected himself. "He loved to gamble, but he lost a lot."

Before it all came apart, Natale figured he gave Merlino between fifty and sixty thousand dollars from the meth business. It was a minor share in a much larger operation that stretched from South Philadelphia to the South Jersey shore.

Constantine, who lived in Sea Isle City, New Jersey, was storing P2P in a small garage he rented outside of town. He, in turn, got his own brother involved in several deals. It was a real family affair. Previte, working for the FBI and using money provided by the feds, made half a dozen "buys" from Rubeo, Santilli, and Constantine. Previte's deals were only a small part of a bigger operation, but they provided the FBI with the evidence it needed to prosecute Natale's drug ring.

Natale never appeared to worry about whom he was putting in jeopardy. He wanted and needed money. He had been away for fifteen years. Now he was the boss of the mob and he was going to live the life of a Mafia don—live it the way he had always imagined it would be.

He and his wife had moved from a small apartment to one of the penthouse units in a high-rise overlooking the Cooper River in Pennsauken, New Jersey. He needed a car, he needed clothes, and he needed cash to cover his expenses—dinners, trips to the shore, nights out drinking with associates.

Ralph also had a girlfriend. And as anyone who has followed the soap-opera-like romances of mobsters in love—or in lust—knows, a girlfriend costs money.

*   *   *

"They're gun molls, there's no other way to describe them," says Previte. "It's the action. They crave the action. They want to be with you because you're a gangster, because they think you're dangerous. . . . I had a good wife. She left me because I was a fuckin' bum. I've had a lot of good sex since then, but not a good woman."

Previte has thought a lot about the mob and women, and he has several different explanations for how and why he was able to get so much sex.

Sometimes, he would say with a twinkle in his eye, it was because he was so charming. "I always made them laugh," he said. "I entertained them. A woman wants to be entertained. Guys don't realize that. I wasn't the best looking, but I always went home with the broad. The other guys, they went home with the *Daily News*."

But on other occasions, Previte would say that romance in the underworld was a power trip, a cash-and-carry, what-can-you-do-for-me proposition.

"Gangsters carry what they got in their pocket," Previte said. It was not a sexual reference. "Every time you reach in your pocket you're pulling out five Gs or ten Gs. I had this girl, used to come over to see me. Beautiful. She was married. I'd get her some coke and we'd party. I was screwing her for fifteen years. She wasn't coming over 'cause I looked like Tom Cruise.

"It's the action, the excitement. The danger.

"A wiseguy, generally, he's not a nine-to-five guy. He's got a wife, they have kids. She takes care of the house. But he's still out partying. She gets older. She's got a little cellulite. He's still gonna get young broads. What's the attraction? What's not the attraction? I was getting pussy seven days a week. It was like a job. I needed to get a social director. . . . Look, a regular woman, even if she was attracted to me, would drive a hundred miles an hour to get away from me. They're gun molls. I wouldn't want my daughters going out with somebody like me.

"It also cost me a lot of money—a *lot* of money. But at the time, I had it. . . . You got that big knot in your pocket. If I ever figured it out, I probably spent a half-million dollars on broads. . . . You pay for these big fuckin' dinners. You're out at a club drinkin' till two in the morning. You're buying the broad whiskey, or getting her coke. You're spending money on hotel rooms 'cause you can't take 'em home.

"When I was in the police department we used to call it 'big wheeling.' The woman at home wants a nice house. These gun molls want rings and jewelry and a good time. It all costs ya.

"What's the attraction? What's *not* the attraction? They were young, good-looking, with hard asses and no cellulite."

Ralph Natale had a wife and five grown children. His wife, Lucia, had stood by him for the fifteen years he was away in jail. The fact that none of Natale's sons or daughters ever got in serious trouble with the law is a tribute to Lucy Natale, not Ralph. He was gone for a good part of their lives. And when he was around, he was hardly a role model. Honesty and integrity were not his strong points.

When Natale returned home from prison, both he and his wife were approaching sixty. Ralph had needs that Lucy apparently was unable to satisfy. On Thanksgiving of that first year, Natale's youngest daughter, Vanessa, brought one of her girlfriends home to celebrate the holiday. She was twenty-five. From South Philadelphia. Her name was Ruthann Seccio.

Within weeks, she and Ralph Natale were "dating." For the next three years, she would be Natale's girlfriend. He lavished her with gifts, clothes, and jewelry. He would set her up in an apartment in a classy development known as Main Street in upscale Voorhees, New Jersey. He rented a home at the Jersey shore for her each summer. He gave her twenty-five hundred dollars a week as an "allowance" and arranged for an associate to lease her a car—a Cadillac. There was nothing that Ruthann wanted or needed that Ralph wouldn't get for her.

She, in turn, spent her free time—and there was a lot of it—on his arm, a cute, young, well-built blonde whom Natale loved to show off. They were in love, she would later tell the Philadelphia *Daily News*. He was going to spend the rest of his life with her.

Stories about wiseguys juggling girlfriends and wives are nothing new. One member of the Merlino crew took up with a twenty-five-year-old and gave her a five-thousand-dollar engagement ring even though he was married and had kids at home. He got an apartment in the city and spent two or three nights a week there "working late."

But Natale pushed it even further.

He and Ruthann went to restaurants and bars together. She accompanied him to mob parties. The girlfriends of other wiseguys never seemed to mind. But several wives, who occasionally appeared with their husbands at affairs Ralph attended with Ruthann, were more than a little upset.

Maybe they knew their husbands were running too. Maybe they didn't. What they did know is, you don't flaunt it. You don't bring your *comare* to an affair when other members of the organization are bringing their wives.

Natale's behavior with Ruthann may seem like a minor detail, but to Previte it was another example of the mob coming apart at the seams. The old rules were being ignored. It was a time of celebrity and self-gratification. Ralph Natale, even as he turned sixty, was trying desperately to keep up with the new generation.

"A fuckin' gun moll," Previte called Ruthann.

"One of my guys had a tanning salon. She used to come in all the time to get tan. Never made an appointment. Never thought she had to. Never paid. That's just the way it was."

Ruthann Seccio became a Mafia princess.

And a source of friction between Natale and many of the members of his organization, especially the younger crew that reported to Merlino. Most of them knew "Ruthie" from the neighborhood. Now, as Natale would tell anyone who mistakenly referred to her that way, she was "Ruthann." It was clear from his tone that he expected her to be treated as someone special.

George Borgesi was one who fell out of favor with Natale in part because he failed to show Ruthann the respect Natale felt she deserved. "She was a broad from the corner," Borgesi said in disgust. "And he wanted us to treat her like Princess Di."

It was petty nonsense, like the dating rituals of high school. But then most members of the Merlino-Natale organization had never really gotten off the corner. Now they were the mob—La Cosa Nostra, or at least what passed for it in Philadelphia at the end of the century. Mentally, however, they were juveniles—juveniles with guns and attitude.

It was a deadly combination.

One of the first to pay the price was Michael "Dutchie" Avicolli.

On April 3, 1996, Avicolli was reported missing. At the time, he was living with a cousin, having separated from his wife because of what police would later describe as "marital problems" at home. Those problems are the key to his story.

Avicolli, who was forty-four, had been around the mob all his life. Dark and handsome, a South Philadelphia lothario, he was one of those guys who never had trouble attracting women. Even when he wasn't trying, broads came on to him. During the Scarfo era he took a beating because an older wiseguy was upset that a woman he had taken a liking to was enamored of the then younger Dutchie.

"A good-looking kid," said Nick Caramandi, the informant from the Scarfo family. "And a tough kid. He wasn't afraid."

Avicolli, described as a "high-ranking associate of Merlino's" during the Stanfa war, was one of nearly a dozen mobsters picked up on gun possession charges during the police crackdown back in 1993. The charges against nearly everyone else found with a weapon were eventually dropped, but the feds took up the gun case against Dutchie. It was an attempt to squeeze him, to try to get him to cooperate.

He refused.

Avicolli had been convicted twice for minor narcotics offenses and once for burglary in the early 1970s. As a result, he could have been sentenced to fifteen years if found guilty of the federal charge of possession of a weapon by a convicted felon.

Cops found a loaded handgun in Avicolli's Cadillac when he was stopped on September 18, 1993. He had just left the apartment, in the Old City section of Philadelphia, where Merlino and his then girlfriend, Deborah Wells, were living.

This was in the middle of the big mob war. It was a month after Merlino and Mike Ciancaglini had been shot, and a day after another Merlino associate, Frankie Baldino, was gunned down outside the Melrose Diner in the heart of South Philadelphia.

Like Baldino, Avicolli was on Stanfa's hit list. In fact, Avicolli had been warned by both the feds and the cops that members of the Stanfa organization were gunning for him. He eventually beat the federal gun rap by launching an aggressive defense: His lawyer argued that Avicolli was "justified" in carrying a gun because the warnings of authorities had caused him to fear for his life. The lawyer, Joseph Santaguida (who also represented Merlino), filed a pretrial motion demanding that several Stanfa associates who were then cooperating be produced as witnesses to substantiate the fact that Avicolli was a target.

Santaguida, turning the feds' own words around on them, borrowed lines from Stanfa's racketeering indictment in describing the danger his client faced. This was a mob war, the lawyer wrote, in which members of the Stanfa organization had armed themselves with "explosives, an Uzi submachine gun with silencer, Tech semiautomatic handguns, .45-caliber semiautomatic handguns, 9mm semiautomatic handguns, .38-caliber revolvers, sawed off shotguns, a street-sweeper shotgun, and a 30/30 rifle with scope."

The unregistered .380-caliber pistol police found in Avicolli's car, Santaguida argued, was a weapon of self-defense for a man mourning the death of a friend—Frankie Baldino—and in fear for his own life.

In December 1994, rather than produce the cooperating witnesses and expose them to cross-examination before the Stanfa trial, federal prosecutors dropped the gun charge against Avicolli. Freed from house arrest, he was back on the streets and back in business.

Avicolli ran a little bookmaking and loan-sharking out of a clubhouse near Ninth Street and Moyamensing Avenue in South Philadelphia. Because of his convictions, he could not be seen associating directly with Merlino, who was on parole at the time. But associates of the young mob leader traveled back and forth between Merlino's coffee shop on Passyunk Avenue and Avicolli's clubhouse a few blocks away.

Through 1995 and the early part of 1996, Avicolli was very much a part of the Merlino organization. Having put his life on the line for the young mob leader, he was now in a position to benefit from Merlino's rise to the top of the organization.

But he also had another "relationship" that did not play well in the underworld. After separating from his own wife, Dutchie had begun seeing the wife of a top Merlino associate. Their not-so-secret affair went on for over a year. Eventually the Merlino associate found out.

The day after he was reported missing, police found Avicolli's car parked a few doors from his clubhouse. He had vanished without a trace. His clothes and all his personal belongings were still in his cousin's home. Family members said they had no idea what had happened to him, or where he was.

Over the next several months, however, rumors began to circulate about Dutchie's fate. The story, built on gossip, innuendo, and a South Philadelphia fascination with the Mafia soap opera, had several different spins. The most titillating was the version that had Dutchie taking up with the mob associate's wife because the associate had seduced and/or had an affair with someone in Dutchie's family.

It was *Moonstruck* without the happy ending.

Natale later provided federal authorities with some details concerning what followed. But as with almost everything else Natale had to say about the inner workings of the Merlino faction, he had very little to back up or corroborate his story.

According to Natale, the cuckolded mobster and one of his underworld friends arranged a trip to North Jersey with Avicolli. The ostensible purpose of the trip was to meet an associate of the Genovese crime family who had been lobbying to switch his allegiance and become a member of the Philadelphia organization. Natale had been in prison with the North Jersey gangster but had been putting him off for some time; he wasn't eager to ruffle any feathers with the Genovese organization, the largest and most violent crime family in America.

Only three people know firsthand what happened when Merlino's associate took Dutchie Avicolli to North Jersey that day. Avicolli did not make the return trip to South Philadelphia. According to law enforcement sources, he was shot and buried in the woods. In an apparent attempt to curry favor with the South Philadelphia wiseguys, the Genovese crime family associate already had a grave dug when they arrived for their visit. Avicolli's body has never been recovered, and no one has been charged in connection with his disappearance.

Natale said he knew nothing about the hit before it took

place. He was told that Avicolli was killed because he was "getting loose-lipped." Natale didn't buy it. If Dutchie kept his mouth shut when he was facing a fifteen-year prison sentence in the gun case, why would he start talking two years later? It didn't make sense. Avicolli was killed, Natale told the FBI, because he "was having an affair" with another mobster's wife.

To this day, Michael Avicolli is listed as a missing person by the Philadelphia police. In the months following his disappearance, members of the Merlino crew would occasionally joke about Dutchie being spotted in Atlantic City or Las Vegas. But neither his former mob friends nor the police investigating his disappearance expected to see Avicolli again.

Six weeks after Dutchie went missing, Anthony Milicia was gunned down on a South Philadelphia street corner. There has never been any doubt about who was behind that shooting, or why it happened.

On the morning of May 29, 1996, Milicia, who was known as "Tony Machines," was shot as he sat behind the wheel of his SUV near the corner of Thirteenth and Dickinson Streets.

Milicia was seventy-six years old. He was a partner in a lucrative vending machine company that had close to a hundred illegal video poker machines in bars and restaurants throughout South Philadelphia. More important, he and his partner, Louis "Pegleg" Procacinni, were serious and wealthy businessmen.

Always operating on the fringe of the mob, but never a part of organized crime, they had survived and prospered despite attempts, first by the Scarfo organization and then by the Stanfa crew, to get their hooks into the business.

This time they wouldn't fare as well.

Natale had had his eye on the Milicia operation for more than a year. He wanted half, which he estimated could amount to more than a hundred thousand dollars, tax-free, per year.

"First of all, they had a huge lottery business," he said. "Then they had an even greater poker machine business."

Milicia and his partner had become quasi-legitimate mortgage brokers. The business went like this: If you wanted to open a bar in South Philadelphia and needed money to get started—cash to buy the property, renovate the building, whatever—Tony Machines and Pegleg would front you the cash. You'd have to pay the money back over time, just like any mortgage. In many cases, there were signed bank notes. It was a legitimate loan against the property.

But the key to the deal was this: you had to agree to allow Milicia to place his machines in your bar or restaurant.

"Poker machines make more money . . . than even selling drugs," Natale said in awe as he described Milicia's entrepreneurial skills. "They're in almost every bar in South Philadelphia. . . . They're in laundromats. They're in delis. The profit on these poker machines is enormous."

And after the initial cost—a machine could be purchased for about fifteen hundred dollars—the only expenses were routine maintenance and collecting each week.

Milicia and his partner took half of what any machine generated until the bar or restaurant owner paid off the loan. After that, a different arrangement—one that was more beneficial to the bar owner—might be worked out.

"In even a small place," Natale said, "let's say the take is eight hundred dollars a week. That's profit because you can't lose with a poker machine. It's set up that way. . . . You give him four hundred, you keep four hundred. Imagine having fifty or sixty of those machines. That's every week. That's fifteen thousand to twenty thousand clear a week. . . . That's big money."

Ralph Natale kept imagining. Through most of 1995, he and Merlino tried to move on Milicia and his partner.

"In every other city," Natale complained, "everything illegal belongs to La Cosa Nostra. . . . All the poker machines belong to La Cosa Nostra. I came home. It belonged to these two guys. They were making . . . millions of dollars every year on poker machines. That's clear, cash money."

But Milicia kept ducking Natale. He wouldn't show up for meetings; he moved around at odd hours. Procacinni met with Natale on several occasions to try to smooth things out, complaining that his partner, Milicia, was a "hardhead," but also claiming that Milicia was "afraid" to meet with Natale.

At Christmastime that first year, Milicia and Procacinni came up with a "package," an envelope stuffed with thirty grand, Natale said. They also agreed to "go partners" with Merlino and Natale in their numbers business. But Natale never saw any money from the lottery. "They kept claiming they were in the red," he said.

He never believed it. He knew a little bit about the numbers business.

"No matter what the number is, you allow for the hits, you allow for paying your percentage of commission to who's your writers in the street, no matter what happens, your profit margin . . . is between ten and thirteen percent. No matter what happens."

Natale figured Milicia and his partner were writing about a hundred thousand dollars a week in numbers. He wasn't seeing a cent, even after they took him in as a partner.

Natale was able to wrestle another fifty grand out of them by negotiating a "loan" during a breakfast meeting with Procacinni at a diner in Brooklawn, New Jersey. Pegleg came up with the money, even though they both knew it would never be repaid. Natale said he kept twenty-five thousand, gave Merlino twenty, and gave five to Steve Mazzone.

At the meeting Natale also informed Pegleg that the mob was going into the video poker machine business, and that sooner or later Milicia would have to sit down and "settle up."

Natale, seething over the lack of cash from the numbers

business and angry that Milicia refused to come to any meeting, decided it was time to send a message. "Tony Milicia kept avoiding the issue," Natale said. "He would not meet with me or with anyone else I would send."

Natale met with Merlino. "We're being made assholes," he told the younger gangster. "We gotta get rid of this guy."

Just as he had done with the Billy Veasey hit, Natale said he left the planning to Merlino and George Borgesi. As a result, when he later testified about the shooting of Tony Milicia, Natale's information was based on what he had been told.

What is certain is that a gunman, shooting from a passing car, opened fire on Milicia as he sat behind the wheel of his Ford Explorer parked near the corner of Thirteenth and Dickinson Streets in South Philadelphia. Also in the vehicle at the time was one Bonnie Leone, owner of the nearby Bonnie's Capistrano Bar, a bar Milicia had once owned. Milicia was apparently dropping Leone off that morning. The shooting occurred around 9 A.M.

Four or five shots were fired into the Explorer, shattering three windows. Milicia was hit once in the chest and also suffered cuts from the flying glass. Leone was not hit; according to one witness, Milicia forced her down in the front seat when he spotted the gunmen, shielding her from the gunfire. Milicia was rushed to nearby Thomas Jefferson University Hospital in critical condition.

Police said the shooting had all the earmarks of a mob hit. But investigators also noted that Milicia, who carried large sums of cash, could have been the target of a robbery gone awry. He had been robbed several times in the past. Neither Milicia nor Leone provided investigators with any details.

Sitting at the end of the long, narrow bar in the Capistrano a few hours after the shooting, Leone told a reporter, "I don't know anything. I didn't see anything, and I really don't want to say anything." Then she turned away, resuming a conversation with two other patrons.

Against a wall in the corner was a brightly lit video poker machine.

It was a "Dodge City" model.

Ralph Natale said he was told that Joey Merlino was driving the car that day and that Frank Gambino did the shooting. He was angry that his suggestion that they use a shotgun—"That's a final shot," he had said—had been ignored. Milicia survived.

But from a business perspective, the shooting proved effective.

Milicia, who lived in South Jersey, spent weeks in the hospital, and then months holed up in his home in Somerdale, recuperating. Procacinni, meanwhile, sat down with Natale and negotiated a peace.

They met in a restaurant off the old Racetrack Circle in Cherry Hill. Merlino, Borgesi, and Mazzone were also at the table. The mob was about to go big-time into the video poker machine business.

Natale set the terms.

First of all, he said, the mob was putting its own man in place to keep track of the numbers business. Natale said he wanted to see a profit and he wanted money on a monthly basis.

He also told Pegleg Procacinni that his organization was taking over any machines that were in bars and restaurants owned and operated by mob associates. He estimated that would be about twenty of the machines Milicia and Procacinni had on the street.

Finally, he told Procacinni that he and Milicia could not negotiate any "new stops." From here on, the mob would control the distribution and placement of machines in the city. And by the way, Natale said, "we might need a little help with a mechanic to get set up."

It was a sweet deal—the kind of arrangement that Natale had always expected as a perk of being a mob boss. He had been struggling for more than a year to establish himself, trying and failing a dozen times or more to launch legitimate or quasi-legitimate businesses. Now he had one. All it took was one attempted murder.

But he wanted even more. Pegleg and Milicia also "had to pay us a thousand dollars apiece a month, that's separate from anything. One thousand to me and one to Joe Merlino," Natale said.

A few weeks later, Natale—who was sneaking into Philadelphia on a fairly regular basis now, despite his parole prohibitions—met with Merlino, Borgesi, Mazzone, Marty Angelina, John Ciancaglini, Gaeton Lucibello, and Frank Gambino.

Natale told them about the poker machine deal. He also introduced a new financial backer: a local bookmaker named Danny D'Ambrosia was going to front money for the purchase of more machines the organization would place in bars and restaurants. Some of these would be "new stops." Others would be locations where the mob's machines would replace those of independent distributors.

The bar owners would have no choice. If they wanted to stay in business, they would take the mob's machines.

"Go into saloons," Natale said. "Anybody that's got one that's not ours, take them out. Tell them we want them out of there."

Natale told Merlino and the others that they would share equally from whatever income the machines generated. "It would be a comfortable living for all of us," he said. "It's anywhere between a thousand, maybe, to twelve hundred apiece. Every week."

With the additional monthly tribute payment of one thousand dollars from Milicia and Pegleg, Natale and Merlino were each on their way to making six figures from the poker machines alone—tax-free.

About a year later, Natale met Milicia face-to-face for the first time. They were in Gino's Café, a small neighborhood bar that Borgesi had taken over early in 1997. Milicia and Pegleg were holding a mortgage on the place for the former owner, but Natale let them know that Borgesi would be in no hurry to pay it off.

He and Milicia had a few drinks that night and seemed to get along; after all, Milicia was closer in age to Natale than Natale was to most of the young mobsters.

"He was very happy to do what he had to do," Natale said of Tony Machines. "And really, in a way, he wasn't a bad fella."

Natale told Milicia he was "glad" Milicia had survived the shooting. Milicia nodded and smiled. Given the alternative, he was happy to be there.

Gino's Café quickly became a mob hangout, a place to go for drinks and a nice sandwich—sliced pork, roast beef, meatballs, maybe a little broccoli rabe on the side. That summer Gino's sponsored a softball team in a local neighborhood league. It was unlike any other softball team ever assembled outside of prison walls.

Joey Merlino was the pitcher. George Borgesi played third base. Steve Mazzone was the shortstop. Borgesi's uncle, Joe Ligambi, who was released from prison in February 1997 after his conviction for a 1985 gangland murder was overturned, played second base. Angelo Lutz was the manager. Only in Philadelphia did the mob have its own softball team.

Gino's Café attracted a large following. The players wore black baseball caps, black T-shirts, black shorts, and black high-top Nike sneakers. The shirts, with the words "Gino's Café" scrawled in white, quickly became sought-after collectors' items.

The funny thing was, they could play ball. They won more often than they lost. Sometimes they won big. After one blowout, someone suggested they should call themselves "the hit men."

Not everyone laughed.

At most games, the FBI and the Philadelphia Police Department had cameramen in the stands. It was a chance to get live shots of mobsters in action. Merlino, finally off parole and free to associate with whomever he chose, seemed to love the attention.

Borgesi, on the other hand, would bristle.

"All we're doing is playing ball," he said. "Why can't they leave us the fuck alone?"

8

The Gino's Café softball team marked the culmination of more than two years of high-profile living on the part of Skinny Joey Merlino. It also perfectly captured the attitude of the young mob leader and those around him: a combination of "fuck you" and "look at me, I'm a gangster."

This came a few years after John Gotti crashed and burned in New York. But the lesson was apparently lost on the Philadelphia wiseguys.

Gotti, the Dapper Don whose name appeared in the gossip columns as often as it did on the metro page, reinvented the role of Mafia boss. He was on the cover of *Time* and *People* magazines. A generous *padrone,* he hosted an annual Fourth of July barbecue and fireworks show for his neighborhood each year.

A media darling—not for nothing was the first book written about him called *Mob Star*—Gotti was ready with a sound bite for the television reporters each time he beat a case in state or federal court. He was the Teflon Don: for a while, it seemed that no charge could stick.

He was the prototype—the celebrity gangster of the eighties and nineties. And a 1992 racketeering conviction that sent him to prison for the rest of his life did little to tarnish that image.

Four years later, Joey Merlino had become the John Gotti of Passyunk Avenue.

It started with the first Christmas party back in November

1995, but it really took off a year later, when Joey threw a party for three hundred and fifty of his closest friends.

It was billed as a bash to celebrate the baptism of his infant daughter, Nicolette. But it was really a coming-out party for the flamboyant young wiseguy.

Nicolette was born in July 1996 to his girlfriend, Deborah Wells. Joey's party was on November 17, two days after he came off parole and was for the first time in two years free to associate with whomever he wished. Friends and family gathered that night at the Benjamin Franklin House, a stately hotel turned condominium complex on the fringe of Philadelphia's historic district.

It was the underworld social event of the year.

While the hotel rooms had been converted into luxurious condominium units, the second-floor ballroom, with its elaborate chandeliers and Victorian-style furnishings, was still rented out for social occasions. Joey Merlino and Deborah chose that setting to celebrate the baptism of their daughter.

To law enforcement, it was another, even bolder attempt by Merlino to flaunt his status and celebrity.

"Joey is going to do what Joey wants to do until or unless someone stamps him out," said one organized crime investigator who was tracking the rise of Merlino at that time.

Police and FBI set up surveillance outside the hotel. So had most of the media. Television cameramen and newspaper photographers were swarming all over the corner of Ninth and Chestnut Streets, at the ornate entrance to the Ben Franklin House.

It was like an opening night on Broadway. Cars pulled up one after another, and couples out of *GQ* and *Vogue* would emerge and head for the doors. The air was filled with the machine-gun fire of cameras snapping away. And then onlookers, each with his own motive, would scramble to find out who it was who had just entered the building.

Merlino arrived in a Mercedes, dressed in a neatly tailored

dark suit, highly starched white shirt, and floral tie. His girlfriend, Deborah, was dressed casually, but would change into
a black sheath dress before the festivities began. Nicolette arrived under a white blanket in an infant carrier that her father
gently lifted from the car. A valet got the pink diaper bag.

Only those on an approved guest list got into the party.
Everyone, of course, brought an envelope. On its face, the
money was for little Nicolette. Most of the guests, however,
understood it to be a pre-Christmas present for Skinny Joey.

Ralph Natale couldn't attend because he was still on parole. But nearly every other member of the organization was
there. A contingent from North Jersey, and another from
New York, also came down to show their respect.

"It was a party, that's all," Previte said. "The FBI wanted
to know what was going to go on. I told them, 'Sometimes
a party is just a party.' It was really about giving money to
Joey. I brought five hundred in an envelope. I doubt if his kid
ever saw any of it."

There was a certain irony in the gift-giving for Previte. At
the time, he was running a sports book in South Philadelphia
that was taking heavy action from Merlino. Before the football season ended, Merlino—who often bet through prox-
ies—owed Previte's book $212,000.

"Joey didn't wanna pay," Previte said. "He never paid. He
claimed the bets were some other guy's, and he was going to
collect them. We finally had a sit-down with Ralph. I had to
act upset. I really wasn't, because the FBI was backing the
book, and really we didn't lose any money. It's not like he
ever won. He just wasn't paying what he owed.

"We had a meeting over at Pal Joey's, the restaurant
Ralph was trying to open up in Cherry Hill. And Ralph tells
Joey he can't be doing this to one of our people, meaning
me. Joey insists it's not him, but promises to get the money.
It was all bullshit. He never paid. But that's the way he was.
He was a lousy gambler, couldn't pick a winner. Played all
these combinations. But that was Joey."

None of that mattered on the night of the party, however. It was just seven hours of food and music and more food.

The buffet started with hot hors d'oeuvres and champagne. Then, at different food stations around the sprawling ballroom, guests had their choice of pasta, calamari, scampi, king crab legs, filet mignon, veal chops, and porchetta. There were several well-stocked bars, two orchestras, a string band, and popular disc jockey Jerry Blavat playing oldies when the bands were on break. Before the party ended, there was a breakfast spread of eggs, bacon, sausage, and muffins, which guests could wash down with coffee, orange juice, or champagne.

Ten days later, Merlino would host his second annual Christmas dinner for the homeless, again attracting media attention and enjoying his time in the spotlight. This time the gifts for the kids included brand-new bicycles. More than thirty were lined up along a wall next to the Christmas tree as the homeless kids walked into the party. Their eyes lit up. Their faces broke into broad smiles.

Joey had done it again.

A few days after that, he gave one of the first in what would be a series of media interviews, claiming he was unjustly targeted by the FBI and insisting that he was not involved in the murders and conspiracies for which he was being investigated.

"I don't bother nobody," Merlino said. "I don't want nobody to bother me. . . . Every day I get blamed for something."

He threw in digs about the Ruby Ridge and Branch Davidian incidents, two highly publicized examples of FBI ineptitude. Joey, who read two or three papers a day—usually the tabloids out of Philadelphia and New York—loved reading about cases where the government screwed up.

Merlino went on to chide the local FBI about its investigation, claiming that a grand jury had been hearing evidence for four years and still hadn't returned an indictment against him. Customers at his Avenue Café were being routinely stopped by police and FBI when they left his establishment, he alleged. The owners of any bar or restaurant he and his friends frequented were interrogated about whether they were being shaken down by the mob. The FBI, he alleged, tried to get undercover agents posing as waiters into the party at the Ben Franklin. When the caterer he had hired refused to go along, Merlino said, he was told his tax returns would be audited by the IRS.

It was a laundry list of complaints and lamentations, posed by a mob boss who had decided to use the media to send a message to law enforcement.

If the feds have a case, he said, "tell them to bring it on." If not, "then leave me the fuck alone."

Jim Maher, head of the FBI's Organized Crime Unit at the time and the guy running the Previte undercover operation, made no secret of how he felt about Merlino, his celebrity status, and his decision to speak out.

"Maybe he thinks he's Al Capone or John Gotti," Maher said a few weeks after the Ben Franklin bash and the Christmas party. "Capone used to feed the homeless in Chicago. Gotti used to have those big Fourth of July celebrations in New York. Look where they ended up. Capone died of syphilis and Gotti's gonna die in jail."

Just who Joey Merlino thought he was is a question that would puzzle and intrigue the media and frustrate and annoy law enforcement for several more years. Even members of his own organization would occasionally roll their eyes and wonder.

Merlino was one of a kind, a South Philadelphia original. While he clearly had the arrogance and swagger of a Gotti or a Scarfo, there was something else about him that softened the edges: a sense that Merlino was laughing at himself, that he almost couldn't believe his own status or the attention he attracted from police and press alike. "Can you believe this shit?" he would ask friends and associates, shaking his head and smiling.

There was, for example, a moment in the midst of the gang wars and the investigations and the underworld threats that had become so much a part of his life, when he was asked by a television reporter about a five-hundred-thousand-dollar contract that some mob leaders had supposedly put on his head.

By that point, in the late 1990s, Merlino was unfazed by the constant media attention. Like Gotti, he had become a master of the one-liner, the quip, the self-deprecating aside. He seemed to enjoy the give-and-take. He had a rapid-fire delivery, his words spilling out at a staccato clip. Whether he was talking about the Phillies, the Flyers, or the latest FBI investigation, Skinny Joey shot from the hip.

But at this particular moment he paused for a second to think about the question. Then a crooked smile creased his handsome face, and his dark eyes flashed.

"Gimme the half million," he said, "and I'll shoot myself."

Joey Merlino, of course, had been born into La Cosa Nostra. His father and uncle were ranking members of the Scarfo crime family. The mob had always been part of his life. But so were the street corners of South Philadelphia, a section of the city defined by its tightly knit ethnic neighborhoods and century-old parish churches. And they, as much as the Mafia, had left a clear imprint on his personality. Merlino was a younger, hipper brand of celebrity gangster, one whose lifestyle was less *GQ* than *Details* or *Maxim.* Stylish and good-looking, he was a fixture in the clubs, bars, and restaurants that catered to the upwardly mobile and economically flush thirtysomething crowd. His swagger, his cockeyed grin, his South Philly who-gives-a-damn-let's-party attitude took the mob out of the shadows and put it into the gossip columns in Philadelphia as surely as Gotti had done in New York. He was a rock-and-roll gangster whose egocentric sense of entitlement was the underworld reflection of the me-first style of his contemporaries who were making it at that same time on Wall Street, along Madison Avenue, and in Silicon Valley. Young and self-assured, they all were riding the boom.

None anticipated the bust.

"You gotta understand about Joey," an associate once said. "If he's got five grand in his pocket, he's gonna spend it. Because tomorrow he figures he's gonna go out and get five more."

"Money don't mean nothing to him," said another. "All he cares about is bettin' and broads."

"Joey was a guy who danced to his own beat," said Frank Friel, formerly a top organized crime investigator with the Philadelphia Police Department. "He epitomized the new La Cosa Nostra. He enjoyed the trappings."

Friel predicted that there were only two things that could put a crimp in Merlino's style—a conviction or a bullet. For years, and against all odds, Merlino ducked both.

"You know in this business, it's good to be smart, but sometimes it's better to be lucky," said another wiseguy who followed Joey Merlino's career. "Joey was always lucky."

Not in the conventional sense, however. From the time he was a teenager, Joey was a degenerate gambler. He loved to bet, especially on sports. But he wasn't much of a handicapper, so he often had to scramble to cover his losses, and he was always looking for a way to make money.

One of his bolder gambits—and a move that resulted in his first serious jail time—came in 1987, at a moment when his father, Scarfo, and the others were in prison awaiting trial on racketeering charges. Twenty-five-year-old Joey hooked up with two other guys from the neighborhood and planned an armored truck heist. One of the guys worked as a guard on the truck, which collected large sums of cash from businesses in South Philadelphia. The plan was for the guard to drop a bag of cash at a designated area—a trash bin behind a store at Delaware Avenue and Wharton Street. The plan went off without a hitch: Merlino and an associate were waiting there, as scheduled, to pick up the bag. Later, he'd told the guard, they would split the loot three ways.

The take was $357,150.

Later never came.

Instead, there were angry shouting matches, a fistfight, and gunfire. One accomplice, Richie Barone, saw about fifty grand. The other, the guard on the truck who made it all happen, got about thirteen thousand. Merlino kept the rest. Both men eventually testified for the government. Following his 1990 trial, Joey Merlino was sentenced to four years. The money was never recovered. As one federal prosecutor observed, the prison sentence was the price Merlino was willing to pay for the cash.

Yet by the time he went to jail, most of the money was gone. Merlino, friends said, blew it all on gambling, drinking, nice clothes, and beautiful women. His father, in prison, was livid. His mother rolled her eyes.

The incident was also part of a pattern, a way of life that Joey Merlino would embrace. For years he disguised a ruthless streak behind his street-corner charm and charisma. Most of his friends reacted to his antics the way his mother, Rita, did. They would nod, shake their heads, and say, in effect, "That's Joey." Only when they became the targets of his hustle did the act start to wear thin.

The score from the armored truck heist was just part of the cash that financed Merlino's lifestyle. His father, according to Natale and others, had left between five and seven hundred thousand dollars in cash when he went to jail in 1987. Joey was supposed to use that money to take care of his mother and two sisters. Rita Merlino lived in a nice row house in an upscale residential neighborhood in South Philadelphia, which may have been purchased with some of that money. But a big chunk of the cash slipped through Joey's hands, spent primarily on betting and broads. By the time he went to jail in 1990, Joey, then in his late twenties, had gone through nearly a million dollars, according to several insiders who had watched the young wiseguy at work and at play.

For most of his career, Merlino bet heavily on sports, football and hockey in particular. He loved the Flyers, and when he showed up at the Spectrum for a game he would be surrounded by fans asking for autographs. In the spring of 1996, his friendship with Flyers star Eric Lindros hit the gossip columns, as well as a flurry of radio and television reports. The two, it was said, had been seen "hanging out together" at several popular nightspots.

Not so, said Merlino. "He makes seven million dollars," said Merlino admiringly of Lindros. "He don't need this." They had bumped into each other at a club on one or two occasions. Both were young, handsome celebrities who attracted a crowd wherever they went. Merlino also insisted that the unspoken implication that he was somehow using the connection to help him make money betting on hockey games was ludicrous.

"I've been betting *on* the Flyers," he said during another of the team's annual playoff swoons. "They're killing me." That, of course, was difficult to prove. Only Merlino's bookies knew for sure, and most of them weren't talking.

A few who did speak up later on didn't have anything nice to say about the handsome young mobster. The story was basically this. Merlino bet heavily. When he won, he collected. When he lost . . .

"Crime don't pay" was one of Merlino's favorite expressions. And when it came to losing bets, neither did he. His lifestyle—the gambling debts, the extravagant purchases, even the highly publicized Thanksgiving giveaways—was financed with other people's money.

Still, when Joey Merlino was at the top of his game, nobody did it better. He, more than any other gangster, gave the Philadelphia mob a personality. He brought a face and a style to an institution that, like so many others in the City of

Brotherly Love, always operated in the shadow of its bigger and better known New York counterpart.

His world included The Palm, the posh Center City restaurant that catered to the city's celebrities, power brokers, and political players, and the Eighth Floor, a hip nightclub on Delaware Avenue that was packed each weekend with the young and beautiful. Summers, of course, were spent at the Jersey shore, where he would rent a condo in Margate, party every weekend at the clubs there, and, despite being on the exclusion list, slip into one of the Atlantic City casinos for a night of craps or blackjack.

Life was good for Joey Merlino.

Most of the time he traveled with an entourage, six or seven other young wiseguys and wannabes, all draped in Armani or Versace, their thick hair neatly coiffed, their nails buffed and manicured. They'd arrive in a Mercedes, a Cadillac, or maybe a customized four-wheel-drive Jeep or Land Rover. Valet parking, of course, and then the group would glide toward the entrance of the club. Lines would part. Cover charges would be waived. They'd be escorted to a prominent table, or given the best seats at the bar. Waiters and waitresses would scurry to meet or anticipate their every demand.

And then the buzz would start. Stares and nods and whispers:

"It's Joey."

"Skinny Joey."

"Merlino's here."

They had come a long way from Twelfth and Wolf.

Or had they? One night a woman who had grown up on that same South Philadelphia corner with Merlino and some of his top associates was invited out by her old friends. She was in town visiting her parents. She had been away for several years. And she was amazed by what she saw.

They hit five or six bars along Delaware Avenue that night before ending up at an after-hours club. Everywhere, she said, they got preferential treatment.

"We were waved right through the lines," she said. The best tables were reserved in advance or were quickly made available. "It was like they knew we were coming."

The drinks always flowed for Joey. And though the clubs were crowded, no one was allowed to get near their table unless Merlino or someone else in the group nodded a silent approval to the bouncers who had set up a security watch. But that didn't stop dozens of beautiful young women from tossing napkins with their names and phone numbers written on them at Merlino. Some, to underline the point, lifted their shirts and flashed their breasts as well. "I was amazed," the young woman said.

But she was also struck by something else. "Other than the fact that they all had wads of cash in their pockets, these guys acted no differently than when we were growing up. They were still the guys from the corner."

And just as they had when they were hanging on the corner, most of them looked to Merlino to set the agenda. Eventually they would all pay a stiff price for that unthinking street-corner loyalty.

9

He was fifty-eight years old, but he was from another time.

Joe Sodano was a true believer, proud to be a member of the most exclusive men's club in the world. He couldn't understand why anyone would want to change it.

Caught on tape during the Stanfa investigation in 1991, Sodano, a capo in the Newark, New Jersey, branch of the Philadelphia family, made it clear that he wasn't happy with the "new look" of the American Mafia.

"There's no more hidden people," he complained. "Isn't it a good thing—you could be in the Mafia, right, but nobody knows it?" Instead, he said, there were too many showboats, too many celebrities. "They want to come in with a suit and tie and thirty pounds overweight and then pose."

Even John Gotti, a man Sodano said he respected, didn't get it.

"Ego, ego is a dangerous thing," Sodano said as the FBI tape rolled. "He's a good man, John, but he allowed himself to get caught up where . . . he became on the front page of the magazines and newspapers. He made it too easy for them."

By 1996, five years later, both Gotti and John Stanfa, Sodano's boss, were in jail. Ralph Natale and Joey Merlino were in charge of the Stanfa family. Things had changed even more.

But Sodano was stuck in his old-fashioned ways. And eventually it cost him his life.

Sodano had been aligned with Stanfa during the dispute with Merlino, but he was off the streets by the time the shooting started. Indicted in a racketeering case built primarily around his gambling operations, he pleaded guilty and in May 1992 was sentenced to ten years in a New Jersey state prison.

He did a little more than two, winning release around the same time that Natale and Merlino started putting the organization back together.

Sodano went back to Newark and back into business. He picked up right where he left off, running a big-time gambling and loan-sharking operation. He was a major money-maker, always had been. He was big with video poker machines all over North Jersey and in parts of New York City; he even partnered with several New York wiseguys, and people close to him said that he got along a lot better with those guys than with most of the Philadelphia mobsters from his own crime family.

"He was old school when it came to the rules and the regulations," said George Fresolone, a Newark mobster who had been around Sodano for years and whose undercover work for the New Jersey State Police led to Sodano's indictment. "He knew how things were supposed to be and he believed in them."

Fresolone watched the Merlino-Natale saga unfold from California, where he moved with his family after being given a new identity. The burly former wiseguy, who died of a heart attack in March 2002, was one of the few cooperating witnesses who actually managed to turn his life around, purchasing a professional janitorial business and making a legitimate living on the West Coast, serving movie and television stars, Hollywood writers, and film producers.

Fresolone used to laugh about how he was spending his days in exile, shining the marble floors in Cher's sprawling mansion or cleaning the windows in the home of a

star on the TV series *Friends*. "If they only knew," he
would say.

But Fresolone, an unapologetic gossip, also kept tabs on
his old crime family. He watched in amazement as Merlino
and Natale tried to put things back together. He was not surprised that Sodano balked at falling in line under the new
leadership.

Sodano knew that the boss of the family was entitled to a
piece of everyone's action. In the 1980s, he routinely sent
about four thousand dollars a month down to Nicky Scarfo,
Scarfo's share of the gambling and loan-sharking operations
Sodano was running. When Stanfa took over, Sodano sent
cash down to him each month.

Natale and Merlino expected the same tribute. It never
came.

Sodano also ignored them when they asked him to come
down to Philadelphia, on three different occasions, to meet
and discuss business. It was a clear sign of disrespect.
Everyone in the underworld knew the consequences.

Several years earlier, John Gotti had explained to an associate why he had ordered the murder of one of his crime
family members. "You know why he's dyin'?" Gotti asked.
"He's gonna die because he refused to come when I called.
He didn't do nothing else wrong."

Now Sodano wasn't coming when he was called.

"Joey always made a lot of money, but he didn't like to
share," Fresolone said. "That could have been part of the
problem. But I think it was more than that. He was from the
old school. And he hung with a lot of New York guys who
were the same way. I doubt that they respected or recognized
Merlino or Natale. Understand what I mean? If you're true
Cosa Nostra, what do you think about Philadelphia? I mean,
how did Natale and Merlino take over? Merlino was a soldier. What authority did he have to make Ralph? Then Ralph
declares himself the boss? Who ever heard of that? Joey [So-
dano] probably didn't recognize these guys. To him, they

were imposters. That's probably why he didn't come when they called."

And they called repeatedly.

Through the spring and summer of 1996, Natale sent messages to Newark asking Sodano and the rest of the crew to come down for a meeting. Everyone but Sodano responded. His silence was his death warrant.

After becoming a government witness, Natale provided an account of how and why Sodano was killed.

Natale said he and Merlino wanted a piece of the North Jersey rackets. As the boss and the underboss of the family, he said, they were entitled to what they knew was a substantial amount of cash. "They had poker machines, gambling, bookmaking, lottery," he said. "Those proceeds belonged to the Philadelphia family, and we were getting nothing from there. We both knew that Joe Sodano gave a lot of money to John Stanfa."

Natale said he arranged for a meeting at Poppy's, a popular family restaurant in South Philadelphia. They reserved the banquet room and put out a spread. Sodano, Joe "Scoops" Licata, Vincent "Beeps" Centorino, and Peter "Pete the Crumb" Caprio, the made guys in the Newark crew, were called down. Each was expected to bring along two or three associates for the sit-down.

"We wanted them, what we called, to *register* with us what they were doing," Natale said. "Do you have loansharking? Do you have lottery? Do you have the day number? The night number? We wanted to know what kind of business they were doing under the La Cosa Nostra name."

And they—Natale and Merlino—wanted a percentage.

"We weren't receiving one dime from the northern part of the state there," Natale said. "We were supposed to receive at least half."

That may have been wishful thinking: even in the best of times, the North Jersey crew never sent down half. Mob-

sters, particularly guys like Sodano, were financial wizards.
They kept two or three sets of books, constantly cooking the
numbers. Nobody ever knew how much money Sodano was
earning. He kept the real figures in his head.

The share he sent to Philadelphia was usually based in
part on how much he liked, respected, or feared the reigning
boss. And how much he thought he could get away with. At
no time was it ever half of his earnings.

When it came to Natale and Merlino, however, Sodano
sent nothing. And that was unacceptable.

It was Tony Milicia all over again—only this time it was
a made guy who was showing the disrespect. And this time,
Natale knew, he had to make an example that would clearly
resonate throughout the underworld.

Natale set up a private meeting with Merlino at a row
house in South Philadelphia. The two met like that from
time to time: Merlino would pick the spot, and someone
would drive Natale over. Later, when the mob boss turned
government witness, he was never able to pinpoint where
the meetings had taken place. He was told, for example, that
one of the homes belonged to Merlino's grandmother. It
didn't. In fact, Merlino's grandmother had passed away a
few years earlier, and the house had been sold to an Asian-
American immigrant couple. When Natale testified that the
two had met at her house, defense attorneys ripped him apart
on the stand, producing mortgage and deed records showing
that Merlino had no access to the property at the time Natale
claimed.

Whatever the case, Merlino and Natale certainly met in
various mysterious row houses throughout South Philadelphia—just another example of Skinny Joey and his associates keeping Ralph in the dark as much as possible.

In November 1996, after Sodano failed to show up at
three different meetings, Natale said he ordered him killed.
He said he told Merlino to assign the hit to Pete the Crumb
Caprio, who was part of Sodano's crew.

"They're making assholes out of us," Natale told Merlino, "especially that guy [Sodano]. . . . We gotta kill him. Send the message up there to Pete. I want it done. And I want it done this year. Now."

Natale said he wanted to "set a precedent in the Newark area that Philadelphia La Cosa Nostra was back and we were back strong." If that meant killing one of his own members, and a major moneymaker at that, then so be it, Natale said. "Anybody that went against us, even our own people, are going to be killed," Natale told Merlino. "Make sure he's left in the street somewhere."

Natale said it was important that police find the body.

"We'll leave a message to the people up there, the other families from New York and for anybody else up there."

Natale said he assigned the hit to Caprio because he knew Caprio was a "serious" individual. That was Natale's way of saying that Caprio, who was in his late sixties, had committed murder before. Caprio had been involved in several gangland hits. But his reputation throughout the underworld had nothing to do with his prowess as a killer.

"He was a slob," said Fresolone. "That's why he was called 'the Crumb.' He had this little clubhouse on Hudson Street in Newark. A real dive. I used to hate to go in there, it was so filthy. God forbid I would eat something there.

"He had a room in the back with a cot. He used to sleep there sometimes."

Caprio's Hudson Street social club was the scene of his first piece of work for the mob—the hit that got him his button. The target was a guy named Butchie, a bookmaker who was suspected of skimming from a mob-run gambling operation.

Butchie was lured to the social club for a meeting. For several days prior to the hit, Caprio spent hours in the basement digging a hole. First he cracked through the cement: then, with a shovel, he fashioned a grave four feet deep.

Butchie was sipping a drink at the bar in the social club

when another wiseguy, Caprio's cousin, pumped two bullets into the back of his head. Then the shooter and Caprio carried the body into the basement and dropped it in the grave. Just to be sure, the hit man fired another shot into the lifeless body before he and Caprio poured acid over the remains. They then filled the grave with the dirt that Pete the Crumb had left neatly piled next to the hole. After that, they covered the spot over with cement.

And there Butchie remained for more than a decade. In the late 1980s, however, the city of Newark bought the building as part of an urban renewal project. Caprio panicked. Fearing that what was left of the body might be discovered, he and another wiseguy tried to dig up the remains. It was a messy job. The building had been abandoned and the basement frequently flooded. Bones, and what the FBI would later describe as "deteriorating portions of human flesh," were stuck in a muddy hole in the foul-smelling and squalid basement. Caprio and his associate managed to pull out some leg bones, but couldn't get the rest of the body. Instead, they poured more acid over what was left of Butchie, and again covered the hole and sealed it as best they could with cement. Then they took the leg bones to a dump and buried them.

This, then, was the "serious" hit man Natale chose to take out Joe Sodano.

Caprio, in turn, recruited one of his top associates, a guy named Philip "Philly Fay" Casale. He and Caprio had been involved in at least four other murders. Usually the target was somebody who had money and who had crossed the mob.

Whether the transgression was real, or simply fabricated by Caprio and Casale to justify the murder-robbery that would follow, remains an open question. What is clear is that at least four individuals were killed by Casale on Caprio's orders before Sodano's murder.

The four victims were each shot at close range; three as

they sat behind the wheels of their cars meeting with Casale. None suspected what was coming. Casale and Caprio were masters of the setup, cheap-shot artists. Both would later become government witnesses and offer graphically brutal accounts of that classic underworld move—the double cross that ends in murder.

Cold and callous, they joked about each hit. In 1991, for example, Casale was lying in wait for his target. They were supposed to meet in a deserted parking lot in Newark. Bored, Caprio started eating potato chips. He had several bags in the car with him that night. After polishing off a fourth bag, and with still no sign of the target, Casale called Caprio and asked what he should do.

"Eat more chips," Pete the Crumb replied.

Two years later, Caprio tapped Casale for another murder. This time the target was a guy named Billy Shear, a mob associate who was suspected of cooperating with authorities. Caprio described Shear as a "cancer," and told Casale something had to be done "to stop it before it spreads."

Then, with a sly smile, he said, "I think it's potato chip time."

A few days later, after Casale had set up a meeting with Shear to discuss a drug deal, Casale told Caprio, "Tonight's potato chip night."

Both hits went off as planned. In each case, the target was relieved of his cash and jewelry. Casale and Caprio split the proceeds, although not always evenly.

Billy Shear had a thousand dollars in his pocket the night Casale popped him, according to an FBI report. Casale was sitting in the passenger seat of Shear's car when he pulled out a gun and shot Shear once in the head at point-blank range. When Shear began to moan, Casale shot him again. Then, with Shear silenced and slumped over the steering wheel, Casale calmly went through his pockets. He later gave Caprio two hundred dollars, claiming it was half of what Shear had on him that night.

In 1994 Casale used the same MO when he popped Willie
Gantz, a Sodano associate suspected of stealing money from
the mob. Sodano was in prison at the time; if he'd been
around, Gantz would most likely still be alive. Casale's
fourth victim was a businessman shot in a warehouse where
Caprio and Casale thought the victim had a stash of more
than a hundred thousand dollars. They never found the
money.

Casale was a physical fitness nut, a well-built five-foot-
ten fifty-year-old who loved to play the role of wiseguy.
Like most mobsters, he had a small-time criminal history,
including a few busts for gambling. But he also had another
item on his rap sheet that should have made him an unlikely
candidate for membership in La Cosa Nostra—at least the
Cosa Nostra that Joe Sodano believed in.

In the late 1970s, Casale had been convicted of sexually
molesting and brutally beating a ten-year-old girl. He left
her body in a garbage-filled lot after having his way with
her. He spent nearly eight years in jail for the crime; yet
when he emerged from prison, instead of being shunned, he
moved right back into Pete Caprio's orbit.

Pete the Crumb and Philly Fay, the slob and the sex of-
fender: that's who Ralph Natale tapped to—in Caprio's own
words—"bang out" Joe Sodano. That's what things had
come to in the American Mafia by the mid-1990s; that's how
far Bruno's Philadelphia family had fallen.

These were the new men of honor.

Casale was nothing if not consistent. Why fool with a formula that worked?

He set up a meeting with Sodano on a Saturday night in
Newark, in the parking lot of a senior citizens' complex in
the 700 block of North Sixth Street. That's where the police
found Joe Sodano on December 7, 1996, responding to a
911 call at eleven-thirty that night. The building security
guard had received a complaint about a van blocking the

driveway. The guard went to check it out. He heard the radio playing, saw the windshield wipers working, and knocked on the window of the Chevy Astro van. The man sitting behind the wheel did not respond. When the guard opened the door, he saw blood oozing from the man's head.

Sodano had been shot twice, once in each temple.

Casale later told authorities that he was sitting in the front passenger seat when he fired the first shot into Sodano's head. Then he got out of the van, walked around to the driver's side, opened the door, and shot Sodano again. He took a fanny pouch that Sodano was wearing. There was twelve thousand dollars inside. This time he split the cash evenly with Caprio. But he didn't bother to search his victim's pockets. Police found an additional $5,452.50 there.

Joe Sodano, always a moneymaker, was officially out of business.

When Ralph Natale heard about the hit, he was ecstatic.

He told Merlino to set up "a nice little dinner" for Caprio and some of his people. A few weeks later, Pete the Crumb, Philly Fay, and several other members of the North Jersey crew drove down to South Philadelphia.

They met at a restaurant on Passyunk Avenue. There was hugging and kissing all around. Drinks to celebrate the holiday and welcome in the New Year. And, of course, to acknowledge the work that Caprio and Casale had done.

Natale embraced Caprio when he first arrived and told him how proud he was of him.

"I was just doing my duty," Caprio said modestly. "I did it because you wanted it."

Natale smiled. Here, he thought, was a genuine wiseguy. "I got a surprise for you," he said. "I'm gonna make you the skipper up there. You're going to be the captain."

Caprio was overjoyed. Six grand and a promotion. The hit was paying substantial dividends.

That night Natale also asked Caprio about Casale, about whether he thought he should be made, formally initiated. Caprio told Natale that Casale had been "very helpful in a lot of things" that had to be done up in Newark, an apparent reference to the other murders. But when Natale asked him if he wanted to propose Casale for formal membership— "You want to do something for this young man?" Natale had asked—Caprio said no.

"He's not ready yet," Caprio said.

Then they all sat down to dinner.

Natale was riding high. It had been a very good year. He had taken over a big piece of Tony Milicia's operation, a move that had generated tens of thousands of dollars in cash for him and for the organization. It had also sent a message. Now other bookmakers were coming in with cash. The street tax was generating serious money.

Natale was still struggling with the legitimate deals he was trying to get off the ground, but with the Sodano hit he had the Newark branch of the family in line. Caprio, his guy, was in charge. Another source of steady income had been established.

He also had the meth operation. And he had his hooks into two elected officials who could deliver the city of Camden. The future looked good. Out of jail just two years, Natale sat atop a crime family that was back in business.

At the time, he had no way of knowing how fragile his hold on the organization was. The FBI already had dozens of conversations implicating him in extortion, loan sharking, and gambling.

But the best was yet to come.

For months the FBI had been pressing Ron Previte to take the last step, to move from confidential informant to cooperating witness. Previte had balked, considered his options, and put them off.

But in February 1997, they closed the deal. From that point on, Ron Previte was wired for sound. He would spend the next two years moving in the highest circles of the Philadelphia mob with a body wire strapped to his groin.

It was the one spot, he knew, where his macho associates would never look.

10

The agreement with the FBI was a good one, maybe one of the best ever negotiated by a cooperating witness. Previte, who had been getting about six hundred dollars a week as an informant, was bumped up to nearly nine thousand dollars a month, more than a hundred thousand dollars a year. The feds also provided medical and insurance coverage, gave him a car, and covered most of his living expenses.

In addition, the FBI agreed to underwrite his bookmaking operation. He was told he would have to turn in any winnings, but the government would cover his losses.

Previte laughed to himself. It was another chance to make a score.

"They sent this 'gambling expert' up from Washington, D.C., to talk with me," Previte said. "He was an FBI agent. He knew as much about bookmaking as I did about rocket science. It was a joke. But I went along with it."

Ron Previte knew that the only way to survive in the corrupt and treacherous world in which he lived was to be more corrupt and more treacherous than those he was dealing with. Wearing a wire was a part of that game. It was a dangerous but, in Previte's estimation, necessary survival technique.

For Previte, there were no qualms about being a "rat." This was business—Previte simply saw no future in the underworld.

Merlino, he said, was just out to rob and steal and plunder. "Joey's agenda on Monday was to get to Tuesday," he said.

Merlino had stiffed him during the last football season, and Previte knew he'd probably try to do it again. Merlino was also grabbing money from anyone he could shake down. He had begun to strong-arm a group of local hijackers, demanding part of each tractor-trailer load they smuggled out of the cargo terminals that lined the Delaware River in South Philadelphia.

It was a smash-and-grab operation, nothing sophisticated. In New York, wiseguys about the same age as Merlino were worming their way onto Wall Street, setting up "pump-and-dump" stock schemes that generated tens of thousands of dollars per score. And here was Merlino, taking a hundred ceiling fans off a tractor-trailer somebody else had stolen, then sending two or three associates out onto a South Philadelphia street corner selling the fans for twenty dollars a pop, about a third of what they would cost at Home Depot.

"It was embarrassing," Previte said. "This was the mob?"

Over the next year and a half, Merlino took a bigger and bigger piece of each tractor-trailer load, eventually taking it all. The booty would include toy electric train sets, bicycles, women's sweatsuits, computers, televisions, and a load of baby formula that Previte and the FBI ended up abandoning on the New Jersey Turnpike.

Previte also knew that, despite what Natale said, Merlino would kill him if he ever got the chance. He knew that both his life expectancy and his ability to earn money were diminished in a mob in which Merlino held a leadership position.

Natale, meanwhile, was more talk than action. And like any executive who used bravado to hide his own incompetence, he had begun to surround himself with incompetent yes-men.

Tony Viesti, called "Weasel," was Natale's driver and gofer. Previte saw him as little more than a sneak thief, always looking to grab a dollar.

Ronnie Turchi, another old-timer, was tapped by Natale as the crime family consigliere. But he was two-faced, constantly playing one side against the other. He had straddled the fence during the Merlino-Stanfa war, then nearly got himself killed when he tried to buy his way in as boss of the Philadelphia family after Stanfa was arrested and before Natale had come home from jail. Turchi sent ten thousand dollars to a top-ranking member of the Gambino organization in hopes of winning his backing as boss. When Natale heard about it, he reached out from jail and short-circuited the move. He sent a message home through Stevie Mazzone: if Turchi tried to become boss, he was a dead man.

"I'll kill him," Natale said.

Yet Natale named Turchi to the number-three spot in the organization, a move that Previte said made no sense. Turchi, who was a butcher by trade, "didn't know how to make money," Previte said. "He always had his hand out."

What's more, his son had a nasty gambling habit and a penchant for ducking his debts when he lost. According to FBI files, Turchi was called on the carpet several times to explain why his son had not paid nearly seven thousand dollars he owed to a Natale-backed book. Unlike the money Merlino owed Previte, this was a debt that had to be repaid. It was, after all, money that was supposed to go into Ralph's pocket.

Ron Turchi, who eventually would pay the ultimate price for his fence-straddling, lasted less than a year as consigliere. He was demoted to soldier by Natale, who then tapped Stevie Mazzone for the number-three spot.

Natale was also using Danny Daidone as his middleman in Camden. Daidone, whose mobster brother Albert was serving a life sentence for the 1980 murder of a Philadelphia Roofers Union boss, was a wannabe wiseguy who played on the fringes of the mob and city politics. Albert, whose conviction would later be overturned, had set Danny up with a position in the Atlantic City Bartenders Union before he went away. But Danny had been ousted along with a dozen other union officials when the feds took over the labor organization in the early 1990s.

At the time he hooked up with Natale, Danny Daidone was running a furniture store in Northeast Philadelphia, the store where Previte had used a stolen credit card to furnish his Center City whorehouse.

Daidone also provided the furnishings for an office Natale set up in Pennsauken, for a company that was going to go into the business of collecting medical bills for doctors' offices. The company never got off the ground; Daidone would end up getting stuck for the furniture.

Still, he was happy to be around Ralph.

"He idolized Ralph," said Daidone's ex-wife Annette. "He would do anything he asked him to do. Look where it got him."

Previte saw it all coming, and decided it was time to make his move.

"The first wire I wore I got Tony Viesti on tape asking me to lend Ralph fifty thousand dollars for a marijuana deal," he said. "I told Tony I had no problem giving Ralph the money, but he would have to ask me himself."

Previte wasn't just interested in getting Ralph on tape; he also wanted to protect his investment. "With Viesti you never knew. He mighta just put the money in his pocket."

Previte later learned that Natale got the money somewhere else. He did get Ralph on tape at several racetrack meetings, but the conversations were uneventful. Still, he had taken the final step. He was wired up.

"I never worried about my personal safety," Previte said. "But I knew once I strapped that motherfucker on, my days were limited, and my life was gonna radically change. To this day, there are still times when I wish I hadn't done it.

"And I'll tell ya why. I miss Hammonton. I miss my life there. I miss being with the guys there. Every single day I miss it. . . . After fifty-five years of living the way I wanted, that's the roughest thing of all."

His regrets, he said, have nothing to do with the mob, with Merlino or Natale, or with becoming a witness. "I never had one fuckin' iota of respect for them. They were all no good. From Viesti to Turchi to Ralph to Joey. They were cut-throat—Mexican bank robbers. It was like riding with Viva Zapata. They wanted to rape, kill, and plunder, and then go eat and get drunk. There was no Mafia there. Everybody was fucking each other.

"Ralph's running around borrowing money. He ain't gonna pay nobody back. Joey's robbing everybody's book. . . . I felt, even though I didn't want to leave, this was what I needed to do. . . . When I was a crook, I had integrity. I robbed the right people. I wasn't like them. I worked at what I did. It was a profession. I always thought I was smarter than them, and I decided I had to make a move. The bad outweighed the good."

Previte says now that the two and a half years he wore a wire were "the most exciting time in my life. If somebody told me today I could relive those two and a half years and then I'd die the next day, I'd do it. I had that good a time. I just really enjoyed it."

The first tapes Previte made solidified the drug case against Natale. Between February 7 and March 26, 1997, he recorded a series of conversations in which he picked up meth and delivered cash that had been supplied by the FBI. Natale, Rubeo, Santilli, Robert Constantine, and Constantine's brother Thomas all ended up on FBI audio. Several also made video.

Like the original deal at the diner in Cherry Hill, the feds tracked each transaction. On March 1, Previte met with Rubeo in the R&R Bar in South Philadelphia. A few minutes later, Santilli showed up and joined them.

They talked about the money they owed to "Pop"—a ref-

erence to Natale—and said they had given Viesti twenty
grand for one of the last deals. As Previte got ready to leave,
Santilli pointed toward a brown paper bag he had slipped
under the table at Previte's feet. Previte reached down,
picked up the bag, and left.

Inside was nearly two pounds of meth.

Five days later, in the Admiral Lounge, a dingy go-go bar
on the outskirts of Camden, Rubeo introduced Previte to the
Constantine brothers. They sipped beers and talked while
dancers clad in skimpy G-strings and bikini tops worked the
bar for tips, offering lap dances and hand jobs for cash.

Rubeo told Previte that "the old man" had set the price at
eighty-five hundred dollars a pound for the meth he had
picked up at the R&R. Previte handed Rubeo the first of two
eighty-five-hundred-dollar payments at that meeting, and
promised to deliver the balance shortly.

On March 11, in the parking lot of the sprawling Hamilton
Mall a few miles from Atlantic City, Previte met the Constantine brothers. He gave Robert eighty-five hundred to complete
the March 1 deal. Constantine said he had good news. The old
man "said to drop it down to eight for ya." With that, Robert
pointed toward his brother Tommy, who was sitting in a van
parked nearby. They walked over. Robert Constantine reached
into the van, pulled out a brown paper bag, and handed it to
Previte. Inside were two more pounds of meth.

Over the course of the next two weeks, Previte would
make two additional payments of eight thousand dollars
each to Constantine, would deliver a thousand dollars to Viesti, and would listen as the Weasel complained about
Rubeo, the Constantine brothers, and business in general.

"I was making this stuff when those kids were in diapers,"
the sixty-two-year-old mobster said. With a large laugh, he
told Previte that the Constantine brothers had been high
when they met him at the Hamilton Mall and that they
"needed to wear masks" when they were cooking up the
meth in order to avoid breathing in the fumes.

Meanwhile, Viesti said, Natale was getting angry because the meth market was apparently too flush. Other dealers in Philadelphia, Atlantic City, and Vineland had "flooded the market," he said, driving down the price and scooping up potential customers for the mob's operation.

"Everybody's broke," Viesti complained. "It's dry. . . . That's why [Natale is] pissed off."

Through the spring and summer, Previte would record a series of conversations with Viesti and Natale in which they discussed getting back in the meth business, but it wouldn't be until the end of the year that the crank started flowing again.

In the meantime, Previte continued his day-to-day operations in Hammonton. He might go days, even weeks without strapping on the Nagra recording device, then get wired up three or four days in a row. It all depended on what was going on, whom he would be seeing, and where the meeting would take place.

There were times when Previte and his FBI handlers figured it was smarter not to wire him up. In March 1997, Ralph Natale planned a big party for the organization. It was to be held in the banquet room on the second floor of a South Philadelphia neighborhood restaurant called Gianna's, down around Seventh and Wharton Streets.

The party was set for March 16. Both Joey and Ralph had celebrated birthdays earlier that month—Ralph turning sixty on March 6, Joey thirty-five on March 13. Merlino, in fact, had just returned from a short honeymoon trip to Jamaica. He and his longtime girlfriend, Deborah, had gotten married in a small, private ceremony at Saint Monica's Church in South Philadelphia and then took off for some time in the sun.

Merlino had been back for a little more than a week when the party took place. Wiseguys began arriving around 5 P.M. They met upstairs in the banquet room for drinks and a strategy session, in which Natale outlined his plans for the future.

Natale loved to hold court, loved taking center stage. "He was like a football coach," Borgesi later complained. "He was always calling these meetings and giving these pep talks."

Later, around 7 P.M., other guests—lawyers, businessmen, wives, girlfriends—showed up for the actual party.

Gianna's was a piece of the Mediterranean plunked down in the middle of a brick and concrete South Philadelphia neighborhood. The restaurant, only two blocks from the Ninth Street Italian market, featured southern Italian cuisine—pasta, seafood, and veal—in a comfortable and casual setting. But its location in the middle of a block of two-story brick row houses made its stucco facade, tiled windows, and aqua-green awnings seem out of place.

Previte attended the party, but instead of getting himself wired, he had tipped the feds in time for them to bug the second floor. They also had a video surveillance team set up in an unmarked van outside the restaurant, near a cold cuts store on the corner, filming everyone who arrived.

The audio tapes didn't have much investigative value; there was too much noise and the place was too crowded for individual conversations to be picked up in any coherent manner. But the video proved priceless. It was a rogues' gallery of the Philadelphia mob, a valuable piece of evidence for the racketeering case that would follow.

Natale had held other meetings at Gianna's, including one in which he settled a dispute between Merlino and the Pagans, a local motorcycle gang. Merlino eventually became tight with Pagans president Steve Mondevergine, but Natale had been forced to step in because the two were butting heads over money.

Known as "Gorilla," Mondevergine was an ex-cop and dedicated bodybuilder who stood about five foot ten and weighed a brick-solid two hundred and forty pounds. According to Natale, the Pagans wanted to get involved in the bookmaking and numbers businesses, two areas that had always been the domain of the mob.

The Pagans, of course, were used to doing what they pleased when they pleased. They were already big in the meth business; in fact, during the Scarfo years a group of Pagans had clashed with Joey's father, Chucky, in a dispute over money and drugs in South Philadelphia. After two gang members took a shot at the mob underboss during a motorcyle drive-by, Scarfo and the elder Merlino quickly sued for peace. The Scarfo organization may have been built around violence and intimidation, but the wiseguys clearly didn't need to mess with a group that was crazy enough and fearless enough to shoot back.

All of that makes Natale's version of how he settled the dispute between Mondevergine and Merlino somewhat suspect. But as with so much else that Natale brought to the table when he began cooperating, the feds had only his word to go on.

According to an FBI report, Natale agreed to meet with Mondevergine and other members of the Pagans at Gianna's. In anticipation of possible problems, Natale told Steve Mazzone to hide several guns around the second-floor banquet room where the meeting was to take place. "The equipment," Natale said, should be stashed under the buffet table and behind the bar.

Natale also told Mazzone to have two heavily armed associates waiting downstairs. None of the mobsters who attended the meeting were to be armed. But if fighting broke out, the associates were to shoot any Pagans as they came down the stairs.

Natale said that he, Mazzone, Merlino, and John Ciancaglini met that day with about twenty members of the biker gang.

"Natale could see that the Pagans had guns bulging from their waistbands," according to the FBI report of a debriefing session in which he provided details about the sit-down.

But that didn't deter the mob boss from laying down the law. He told Mondevergine and the nineteen other hulking bikers assembled in front of him that all illegal gambling belonged to the mob. The Pagans, Natale said, could do whatever they wanted in the crank business, but bookmaking and numbers were off-limits.

If they didn't like it, Natale said, the Pagans could "go fuck themselves."

With that, the twenty bikers, sitting opposite a sixty-year-old mob boss and his three well-groomed but hardly intimidating associates, acquiesced. At least that's the way Natale said it went down, and the way the FBI reported it. In fact, Natale said, before the meeting was over, Mondevergine asked if Natale could get him a job. Natale said he later arranged for Gorilla to get work as a carpenter at the Philadelphia Convention Center. Natale had a family connection in the union there.

Later the mobsters went for drinks at a bar owned by Gaeton Lucibello. Only then did Mazzone tell Natale that he had never stored the guns in the banquet room as ordered, and had forgotten to assign armed guards to wait downstairs at Gianna's.

Natale told the feds he was "furious" with Mazzone for failing to follow orders. But those who know Natale suspect that he embellished quite a bit in the telling of the story. The final detail—how he unwittingly faced down the Pagans without any firepower in place—is typical Natale braggadocio.

It is also strangely reminiscent of a classic scene from the mob movie *A Bronx Tale,* where a mob boss, played by Chazz Palminteri, tries to negotiate peacefully with a group of bikers who have shown up at his bar. When he asks them to leave, they mock him. Palminteri calmly walks over to the door and locks it. Then he tells the bikers, "Now youse can't leave." With that, a group of bat-wielding mobsters come pouring out of a back room in the establishment and beat the bikers bloody.

Once he began cooperating, Ralph Natale was more than happy to tell the FBI everything he had done—and equally happy, it seems, to throw in a few details with little basis in fact. Whether it was a line from Sammy the Bull or a scene from a movie, all that mattered was that it made him look tougher, smarter, and stronger—a man's man, a true Mafioso.

Previte, on the other hand, gave up only what he thought was necessary. Even after he cut his deal and began wearing a wire, he tried to keep his personal business to himself. And some of it was very personal.

Not everyone Previte strong-armed was an extortion target. Not everyone he beat up was a criminal. There was one guy, for example, whose main offense was that he was giving Previte's then teenage daughter a hard time.

Previte to this day makes an effort to guard his own family's privacy, so he reluctantly provided only sketchy details about the incident. The guy in question was a car salesman. One day Previte's pal Freddy Aldrich showed up at the Honda dealership where he worked and asked to take a car for a test drive. Since Aldrich arrived on foot (Previte had dropped him off), with no car of his own to leave in collateral, the salesman had to go along. They drove several miles, Aldrich complimenting the salesman about the power of the new Honda's engine, the smoothness of the ride, the comfort of the interior.

Before the salesman realized it, though, Aldrich had steered them onto a back road in a desolate wooded area. There was a car, a Cadillac, parked up ahead. Aldrich pulled in behind the vehicle. The salesman looked puzzled. Then, when he saw Previte get out of the car brandishing a gun, he looked frightened—very frightened.

"We put the guy in the trunk of the Honda," Previte said. "He was begging and crying. He thought I was gonna shoot him. It was a blank pistol, but he didn't know that."

Previte fired off a shot and heard the salesman cry out. Then, realizing he hadn't been hit, he began begging again.

"You motherfucker, you better leave my daughter alone," Previte shouted, before firing off another round.

"He got quiet after that," Previte said. "I think he fainted."

With that, Previte and Aldrich got into Previte's Cadillac and drove away. A few minutes later, out on Route 30, the major highway leading from Hammonton to Atlantic City, they stopped at a phone booth. Previte called the auto dealership and asked if they had a salesman by the man's name.

"Yes, we do," came the reply.

"Well, he's locked in the trunk of one of your cars. You better come get him." Previte gave the receptionist at the auto dealership the location, then put down the receiver and drove off with Aldrich.

"The guy never said a word," Previte recalled. "He knew who we were and why we had done it. He also knew we could do it again. Never said a word. And never bothered my daughter again."

Loyalty to family and friends was something Previte believed in.

And he will quickly point out that even though he cooperated with the New Jersey State Police and the FBI for nearly ten years, none of the guys in his core group in Hammonton was ever implicated in a crime because of his testimony. He wouldn't wear a wire on his friends.

There was, however, one exception. And it nearly derailed the undercover operation before it began.

In 1995 and again in 1996, Previte had extorted the owner of a bar in Hammonton, collecting about eight thousand dollars that he shared with Jimmy DeLaurentis, a local police detective and longtime friend. The shakedowns were later detailed in a federal indictment.

The bar owner made the payments to ensure that his establishment—a squat one-story cinderblock building opened primarily on weekends and catering to a largely Mexican migrant worker clientele—did not have its liquor license revoked. DeLaurentis was in a position to pull that plug, and he used Previte to get the message to the bar owner and to collect the cash.

The extortions continued in 1997, after Previte had agreed to wear a wire. The feds had learned about the corrupt cop, and they insisted that Previte tape conversations with the police detective. At first Previte balked. "I liked Jimmy," he said. "And I didn't want to do it. I signed up to make cases against the mob, not the people in Hammonton. I think if it hadn't have been a cop, they would have let it go. But they insisted. I had to agree to it. It's the only thing I did that bothers me. I tried to get him to stay away from me, but he wouldn't. So I did what I had to do."

When his case came to trial, DeLaurentis would claim he was the victim of an FBI sting gone awry—that Previte had manipulated the situation to minimize his own involvement in the extortion, and that he, DeLaurentis, had acted only out of fear of the hulking mobster, who had emerged as the Godfather of Hammonton.

The tapes told a somewhat different story. Sitting over breakfast at a diner in Folsom, just outside of Hammonton, DeLaurentis and Previte discussed the bar owner during a meeting in early March 1997.

"He's gonna pay heavy," the police detective says. "He's gotta go for at least ten."

"How do you want me to go at him this time?" Previte asked.

"He's gotta at least go for ten," DeLaurentis said again. "I been letting him get away with this bullshit."

Previte and DeLaurentis talked for a while about what was going on inside the bar, the scene of frequent fights and suspected drug dealing and prostitution. Previte asked about the crowds and the hookers.

"Sunday nights it's packed," DeLaurentis said.

"Ain't that many Mexicans around now, is there?" Previte asked.

"A lot of them come from Vineland. . . . All come down from Vineland on weekends," the detective said. "You see, they go there because nobody goes in there and bothers them. They go in there and close the door. Nobody bothers 'em."

"They got whores walking outside?" Previte said.

"Sure, by the gate and shit."

"There ain't that many whores working like before, though, is there?"

"Inside, there is. Yeah."

Eventually Previte collected six thousand dollars from the bar owner, in two equal installments. Each time he delivered the cash to DeLaurentis, he was wired. During the second delivery, DeLaurentis balked at taking the money, insisting that they were partners and should split the take evenly.

"That's the other three from the guy," Previte told him as they sat in the diner.

"No, it's yours," DeLaurentis said. "We're working half and half. . . . We always split everything half and half." But he eventually took the money.

DeLaurentis, whose father was the chief of police in the twenty-eight-member Hammonton Police Department, had done other "favors" for Previte, according to federal court documents that became part of the criminal case file. When Tony Viesti was convicted of passing a bad check and sentenced to probation, DeLaurentis provided Viesti's probation officer with a phony letter indicating that he had done fifty hours of community service work for the Hammonton Police Department. And when the Atlantic County Prosecutor's Office, unaware that Previte was an FBI informant, had a surveillance camera secretly mounted on a telephone pole outside a bar where Previte did business, DeLaurentis tipped him off.

"Jimmy was a good cop, real good," said a member of the department at that time. "People liked him. But then he started hanging around Ronnie, and you could see how he

Ron Previte striking a pose during his formative years.

Before his days as a mobster: Previte in his U.S. Air Force uniform.

Before Ron Previte took down the Philadelphia mob as an FBI informant in the late 1990s, it was run by a different set of players in the 1980s—headed by mob boss Nicodemo "Little Nicky" Scarfo *(center)*. Scarfo makes plans with Phil "Crazy Phil" Leonetti *(left)* and Lawrence "Yogi" Merlino *(right)* in Atlantic City at the start of the casino gambling era.

Scarfo *(left)* and associate Phil Leonetti arrive at the Philadelphia International Airport. Scarfo was later convicted of extortion and racketeering and imprisoned.

After Nicky Scarfo was imprisoned, John Stanfa *(right)* became the new mob boss in town. Under the close watch of the FBI, Stanfa and mob associate John Veasey hit the Philadelphia streets.

By the early 1990s, Stanfa's control was threatened by a young and cocky gangster by the name of Joseph "Skinny Joey" Merlino. Here, Joey *(left)* does business with associates Gaeton Lucibello and Michael "Mikey Chang" Ciancaglini.

The rivalry between Merlino and Stanfa soon led to an all-out war. Here, Merlino's men—Stevie Mazzone, Merlino's underboss *(left)*, and associates Gaetano Scafidi and George Borgesi—hang out in their South Philadelphia neighborhood.

A smiling Joey Ciancaglini talks business with his boss, John Stanfa *(left)* and associate Mike Palma in the fall of 1992. Five months later Ciancaglini was gunned down—allegedly by a member of Joey Merlino's crew.

In August 1993, Joey Merlino leaves the funeral of his close associate Mike Ciancaglini, who was murdered in retaliation for the Joey Ciancaglini hit.

Mob boss John Stanfa *(left)* arrives at a grand jury hearing with bodyguards Vince Filipelli and Ron Previte, an FBI informant at the time.

Previte's bullet-riddled silver Cadillac after the ambush on John Stanfa—allegedly orchestrated by Merlino and his men in response to the murder of Mike Ciancaglini.

Ralph Natale and his mistress Ruthann Seccio, his youngest daughter's close friend, in the mid-1990s. Natale and Seccio's relationship became a source of friction in the mob.

After Stanfa was sent to prison, Previte agreed to wear a wire for the FBI and joined Merlino and his gang. From left: Joey Merlino, Stevie Mazzone, Ron Previte, Carmen Perotta, and Michael Lancelotti.

Ron Previte *(left)* and Ralph Natale, mob boss and close ally of Merlino's, at the Sons of Italy Club.

Once Ruthann Seccio's relationship with Natale was

The Avenue Café, the central hangout for Merlino and his crew.

Merlino *(left)*, never one to shy away from the camera, with photographer Brad Nau inside the Avenue Café.

The flashy gangster: Merlino at the height of his criminal career.

Also a fan of the spotlight, George Borgesi, a top Merlino associate and childhood friend *(right)*, smiles for the camera with photographer Brad Nau.

Merlino and his wife, Deborah, at the Benjamin Franklin House. Even the baptism of their daughter was a headline-grabbing affair.

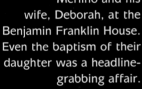

Angelo "Fat Angelo" Lutz, a low-ranking member of Merlino's crew, dresses up as Santa Claus for Merlino's Christmas party at Eladio's Restaurant in 1996.

Angelo Lutz *(left)* with George Borgesi during happier times. Borgesi ranked as the third most powerful player in Merlino's crew, after underboss Stevie Mazzone.

Angelo Lutz *(center)* with defense attorney Joseph Santaguida *(right)* being interviewed by Dave Schratwieser, a television news reporter. Because he lacked a violent criminal history, Lutz was the only defendant allowed out on bail during the 2001 trial.

Michael Lancelotti, a Merlino associate, entering a mob party. Lancelotti was suspected of participating in the Joey Ciancaglini hit but was never charged.

Michael Virgillio, also known as "Mikey Penknife," was a suspected hitman for Merlino.

Joey Merlino *(left)*, Joe Ligambi *(center)*, and Sonny Mazzone (Stevie Mazzone's brother) suit up for the mob's softball team, Gino's Café.

Ligambi exits the federal courthouse in the spring of 2001. After Merlino was convicted and sent to prison, the feds claimed that Ligambi succeeded Merlino as the acting mob boss of the Philadelphia crime family.

was being influenced. We used to wonder, what does he want to be, a cop or a wiseguy?"

Frank Ingemi, a police captain who would succeed De-Laurentis's father as chief, said the relationship changed the department's credibility with other law enforcement agencies. "They didn't trust us," said Ingemi.

The fact that Previte was a mobster was pretty much common knowledge by the mid-1990s, but that didn't stop Jimmy DeLaurentis from continuing his association. He even told some members of the department that Previte was his "cousin."

DeLaurentis started to act like Previte, dress like Previte, walk like Previte. When they met, instead of shaking hands, they would kiss each other on the cheek.

"It was like Jimmy idolized him," said another officer. "Ronnie drove a Cadillac; Jimmy went out and got a Cadillac. Ronnie shaved his head, Jimmy shaved his head. Ronnie started going around in loafers without socks, Jimmy started coming to work in loafers without socks."

From Ingemi's perspective, the low point was the annual Our Lady of Mount Carmel festival in July 1996. The week-long celebration is a combination ethnic and religious holiday, brought to Hammonton a century ago by the Sicilian immigrants who came to work the fields and eventually settled there and came to dominate the community. Ironically, many of those first-generation Italian-Americans were the migrant workers of their day—like the Mexicans who were frequenting the bar that DeLaurentis and Previte were extorting.

Built around the July 16 feast day of Our Lady of Mount Carmel, the festival includes a week-long carnival that attracts thousands of visitors and a feast day parade for which the entire town turns out. Everyone marches, including the police department in full dress uniform. And every year, the members of the department stop by the Sons of Italy Hall for a sandwich and a drink after the parade.

"Usually they sit us in the back, away from the bar," Ingemi said. "Because we're in uniform and it wouldn't look right."

That summer, when he, Jimmy DeLaurentis, and about a dozen other members of the police department were ushered into the back room, Previte was already there with a group of his own guests. Ingemi recognized several of the men Previte was with, including Ralph Natale.

They were all mobsters.

"They said, 'Hi, how ya doin'? Sit down,'" Ingemi said.

There was an awkward pause, and then Ingemi and most of the others made excuses and left.

Jimmy DeLaurentis stayed.

It was just one example of the young cop's fascination with the wiseguys, a fascination that would lead DeLaurentis to "willingly use his badge to orchestrate extortion schemes," as prosecutors later charged, and to "enthusiastically culti-vate his relationship with . . . organized crime figures."

Among other things, authorities would point to two other conversations, recorded by Previte, in which DeLaurentis sounded almost starstruck.

In one, DeLaurentis happily described for Previte a chance encounter he'd had with Natale and several other mobsters at Catelli's, a restaurant in Voorhees Township, New Jersey. Na-tale frequented the spot, which was in the same upscale de-velopment where his girlfriend, Ruthann, had her apartment.

DeLaurentis told Previte he had tried to send over a round of drinks to Natale and the three or four others who were sit-ting with him.

"The waiter came back and says, 'They won't take a drink. Matter of fact, they picked up your bill,'" DeLauren-tis said proudly.

On another, the detective talked about how his father had bumped into Natale at The Pub, a restaurant just outside of Camden where Natale often held court.

"You look awful familiar," DeLaurentis said the mob boss told his father.

"I met you at the Sons of Italy," the elder DeLaurentis replied.

"Yeah, you're right," said Natale.

They spoke for a few minutes. Natale tried to pick up the tab, but DeLaurentis said his father wouldn't have it. He paid his own way. Before he left, Natale said, "Tell your son Jimmy I was askin' about him."

"He didn't ask about me?" Previte said with a laugh.

"No, fuck no," said Jimmy DeLaurentis as the tape rolled.

"Ain't that something," said Previte as he and the police detective laughed again.

Two years later, after he was indicted and arrested, DeLaurentis claimed that he had been entrapped by Previte. In an interview with a Philadelphia television reporter, he claimed he was intimidated by the local mobster, who used the FBI's blessing to run roughshod over Hammonton.

DeLaurentis, pointing to Previte's FBI connections, called his onetime friend "a criminal with a badge."

A federal prosecutor would later use the same description in outlining the bribery and extortion case to a jury in Camden, New Jersey. The prosecutor, however, was talking about Jimmy DeLaurentis.

11

Ron Previte risked his life every time he recorded a conversation for the FBI, but the first real threat that surfaced during his undercover work came not from the wiseguys he had targeted, but from Little Nicky Scarfo, the jailed and maniacal Philadelphia mob boss who was doing a fifty-five-year stint on racketeering and murder charges in the federal prison in Atlanta.

Scarfo was seething over Merlino's rise to the top of the organization. From where he was sitting, Skinny Joey was the "punk" who had tried to kill his son at Dante & Luigi's Restaurant on Halloween night back in 1989. That Merlino had survived and prospered in the midst of the carnage and destruction that had decimated the Philadelphia crime family was a constant irritation to Little Nicky. And as anyone who had followed the career of the volatile mob boss knew, an irritated Little Nicky was a dangerous Little Nicky.

From his prison cell in Georgia, he began to plot a takeover, a way to win back control of the family. His son, naturally, would be his proxy. For that to happen, of course, some people would have to be eliminated. Merlino topped the list. Natale and Previte, the other key figures in the drama that was then playing itself out in the Philadelphia underworld, were also targeted. Using his friendship with Vic Amuso, the jailed Lucchese crime family boss, Scarfo put out contracts on all three. The move, according to an FBI affidavit that surfaced later, was clearly an attempt by Scarfo and his son "to take back control of the Philadelphia LCN family." Nicky Jr. was then living in the Newark area "under the protection" of the Lucchese crime family. The word in underworld and law enforcement circles was that the younger Scarfo had been made—formally initiated into that organization. Little Nicky apparently saw this as an added layer of protection for his son. According to Mafia protocol, killing a made member of a crime family required a sit-down and permission from the organizations involved. That the elder Scarfo saw this as a benefit for his son even as he plotted to kill three made members of his former crime family made little sense, but then there was little about the reign of Scarfo or the tenure of Merlino and Natale that fit in with the traditional values of the Mafia.

The contract was active for several months. Among other things, there was talk of a Lucchese hit team being dispatched to the Jersey shore that summer to ambush a drunken Merlino as he emerged from one of the clubs where he and his associates partied on most weekends. Once again, however, Merlino danced away from danger. The hit never got beyond the planning stages, and Merlino was able to joke on television about the five-hundred-thousand-dollar price tag Scarfo had put on his head. Neither Previte nor Natale ever commented publicly, but the FBI heard a candid discussion in which both laughed and joked about the situation.

Early in 1998, word had started to circulate around the underworld about the three murder contracts. On March 10 of that year Previte, wearing his body wire and a smile, went to the Garden State Racetrack to meet with Viesti and Natale. Everyone was in a good mood. They had celebrated the New Year by pulling off another drug deal and were hoping for even more.

Just before the holidays, Natale had suggested they get back in the meth business. Talking in code, Natale, who was now paranoid about listening devices and convinced the feds had everything wired, told Previte, "Maybe make some Christmas money, sell some Christmas trees." Previte, who had already discussed the matter with Viesti, knew exactly what Natale meant. Viesti had told him that Bobby Constantine, Natale's son-in-law, had begun cooking again, and that a couple pounds of meth would be available shortly.

Two days before Christmas, Bobby's brother, Tommy Constantine, met Previte in the parking lot of a diner in Cherry Hill. He handed him a bag containing two pounds of meth. Constantine told Previte that the price hadn't been set yet, but that "Pop's handlin' this."

Two weeks later, Previte made two seventy-five-hundred-dollar payments—cash supplied by the FBI—to Viesti. In February he learned that Constantine had gotten his hands on more P2P and was preparing to cook up another batch of methamphetamine. It was about that time that the wiseguys also heard about the murder contracts coming out of Atlanta. When he arrived for his meeting at the racetrack restaurant, Previte made light of the report.

"Sure you want me at this table?" he said with a smile.

"I don't know," Ralph said with mock concern. Everyone laughed.

"It means you're recognized," said Viesti, who told Previte he should be proud.

"Ya know, it always makes me feel good," Natale said of the occasional threat. "I've got half a dozen since I got home. . . . Now you really feel like you're part, one of the boys."

"What are you gonna do?" Previte shrugged. He said he'd been through it all before, an obvious reference to the Merlino-Stanfa war. This was no different.

"I appreciate it, cuz," said Natale, trying his best to sound concerned.

They then speculated on where the information was coming from and how it had gotten out. As FBI policy required, the feds had warned all three targets. That was a clear indication that they had picked up the information either through a wiretap or from an informant, Natale said.

"They got rats that are hiding in jailhouses and this and that," he explained.

"In other words, he's so stupid he's talking to somebody that's giving him up," Previte said of Scarfo.

Natale, unaware that he was in that exact position, readily agreed. "That's right," he said. "They talk around one another. . . . It could be coming from, there's a hundred fucking things."

Previte said that Scarfo was "stupid."

"Stupid and vindictive," said Natale. "You never would've liked him."

Previte, playing his role perfectly, complained about the television news report in which Merlino had quipped about killing himself for the half-million-dollar-payoff. The report had implied that Scarfo had sought the approval of the five New York mob families, the so-called Mafia Commission, before putting out the contracts.

Natale reacted immediately. "There is no sitting commission right now," he said as Previte's tape rolled on. "There is nobody reaching out, and the biggest family in New York . . . is the Chin's family." This was a reference to Vincent "the Chin" Gigante, the cagey boss of the Genovese organization who had recently been convicted after ducking prosecution for years by feigning mental illness. The Genoveses, Natale said, had told Scarfo to "go fuck yourself."

"The kid who runs it now told him there's only one boss in Philadelphia, and that's Ralph Natale. And that's where the fuck it's at."

"It's a fucking joke," Previte said of the Scarfo threat.

Natale agreed. But he said it might not be all bad. "It's good to get a little notoriety once in a while," he said. "It makes people—it shakes 'em up. I'm not kidding ya."

From there, Natale and Previte turned philosophical. Both agreed that when your number was up, it was up. It's all written in a book somewhere, they said, and there was nothing you could do about it. So why worry?

"We're all in it the day we're born," Natale said.

"As soon as you're born, the date's there," Previte added.

Who knew if they really believed what they were saying? Each seemed to be playing off the other. Previte for the tapes as much as for Natale. Natale, the showboater, because he couldn't help himself.

But Natale also knew enough to negotiate when in a confrontation with the larger and more powerful New York organizations. Early in his tenure, he had met with members of the Gambino crime family who were interceding on behalf of Joe Stanfa and Joe Ciancaglini. Natale, at the request of the Gambino family, agreed that neither Stanfa nor Ciancaglini would be targeted by his organization again. In exchange, the two had to agree not to engage in any underworld activities. In effect, the crippled Ciancaglini brother and the young son of jailed mob boss John Stanfa were forced into retirement. In the underworld this was known as being "put on the shelf."

Natale, despite his bravado, also tried to work out a deal with Scarfo. At one point he offered, through an Amuso contact, to make Nicky Jr. a capo in the Philadelphia family and put him in charge of Atlantic City. Whether out of distrust or arrogance, the Scarfos rebuffed Natale. He was told Nicky Jr. wasn't interested.

In December 1996, Skinny Joey got out of the coffee shop business, closing the Avenue Café. The place was little more than a hangout and became an easy target for law enforcement surveillance. Merlino could live with that, but he was also smart enough to realize that a routine that placed him in the shop almost every day made him a target of a different sort. During the "situation" with Louie Turra, for example, there were several plots to ambush Merlino in or near the Passyunk Avenue café.

By the time Previte began taping for the FBI, Merlino had relocated. He was now a "salesman" for ARCO Redevelopment, a home remodeling company located on South Broad Street a few blocks from his old coffee shop. ARCO was owned and operated by Albert Coccia, a twenty-eight-year-old guy from the neighborhood who liked being around Joey. Coccia would eventually be jailed on fraud, money-laundering, and cocaine-trafficking charges. The fraud and money-laundering were linked to a scam involving phony home improvement estimates, overbilling, and a ripoff of a federally financed home loan program. Merlino was not tied to any of that. In fact, it did not appear that Skinny Joey did any actual work for ARCO. What he was able to do was collect a weekly paycheck of about a thousand dollars and drive a company car—a Mercedes. The ARCO offices also became a meeting place of sorts for Joey and the guys who were temporarily without a hangout after the Avenue Café shut down.

Previte, angling to get closer to Merlino, stopped in occasionally. He needed to build up the young wiseguy's trust before he could hope to get any incriminating conversations on tape. The fact that Previte was an "earner" got him in the door. Despite the old animosities, Skinny Joey was always looking for another score.

One of Previte's early visits to the office, however, nearly cost him his life.

"I'm in there just shooting the shit with Albert," Previte said. "Joey, Marty Angelina, some other guys are there playing cards. I'm wearing pants and shoes with no socks. I got my feet up on the desk, and I notice the wire is dangling around my ankle. It had come loose and was sitting right out there in the open."

Previte kept talking. For a few minutes he didn't even put his feet down. He didn't want to do anything to call attention to the wire. Then he calmly stood up and said he had to get going.

"The worst thing I coulda done was jump up," he said. "You get in a tight situation like that, you gotta stay calm. My chest is pounding, but I'm not showing any concern."

For Previte, though, on most days, the thrill of the chase far outweighed any concerns he had about the risk he was running. "I just couldn't wait to put it the fuck on," he said. "It was the excitement. I loved it." Only later, when defense attorneys tried to make an issue over the amount of money he was paid by the FBI—the figure would eventually approach one million dollars—did he talk about the danger.

"The money was hazardous duty pay," he said. "Let's see one of them do what I did. What would their lives be worth? I figure I earned every penny."

But if the danger never got to him, the mechanics did.

It took several months before he was comfortable with the device. "First they taped it to my thigh, but the tape was too tight. When they took it off, it was like I had a tattoo, NAGRA, on my leg," he said. "Then I bought a jockstrap and we tried it in there. Then we tried it on my belt, but my belly was too big and we didn't pick up anything."

Finally, Previte got a jockstrap with a cup—the protective plastic device commonly used by catchers on baseball teams. And that's where the listening device would be hidden. Wires would run from the cup through the inside of his pants pocket, where he could control the on-and-off switch.

Previte had wiseguys literally talking to his crotch.

"Every pair of pants I had got wrecked," Previte said. "We cut holes in all of them."

Later, the FBI took away Previte's option of turning the wire on and off. The switch was placed inside the protective cup, out of his reach. The tape was always running.

"They claimed there were technical problems," he said. "I didn't care one way or the other. In fact, it was less for me to worry about. I never was scared wearing it. But physically, it was uncomfortable. If I was wearing it for three hours, I couldn't take a piss. Sometimes it was so hot, I was getting shocks, especially in the summer. I used to wear shorts a lot, so I had to be careful."

One of Previte's earliest tapes was a conversation with Danny D'Ambrosia, a young bookmaker and big-time moneymaker who was being courted by Natale. At around the same time, Merlino and his associates had gotten their hooks into Michael Casolaro, who had one of the most successful sports books in South Philadelphia. Neither bookie was happy about the situation, but their stories provide a vivid account of how Natale and Merlino operated and of how surviving—both financially and physically—often depended on how well you could play the game. D'Ambrosia, when he saw what was coming, figured out how to cut his losses. Casolaro, who thought he could count on a long-standing friendship, wasn't as fortunate. Friendship was always a secondary consideration for Merlino when cash was involved.

"I knew this kid made a lot of money and now Ralph's got him around him," Previte said of D'Ambrosia. "It's obvious what that's about. Ralph sees another source of cash."

D'Ambrosia, thirty-two, was a contemporary of Merlino's. But he brought a decidedly different approach to the business. He would later be indicted for running a million-dollar gambling operation with offices in both Philadelphia and Costa Rica. D'Ambrosia was serious about what he did; he'd been at it, he told Previte, since he was thirteen. While he never discussed actual figures, Previte knew "Danny Dee" was a big-time moneymaker; after all, that was why Natale was trying to bring him under his wing. This was fine with D'Ambrosia, who was aware that Merlino was pushing a street tax and figured an alliance with Natale would save him a lot of headaches. Better to give Ralph an occasional "loan" than to have Merlino and his vultures forcing themselves in as "partners" in his gambling operation.

D'Ambrosia had put up the cash Natale needed to buy

video poker machines after the mob horned in on Tony Milicia's action and was in line to share the take once the machines were up and running. Over time he would "lend" Natale money for other ventures; the tab on those loans eventually grew to more than a hundred thousand dollars. Natale, in turn, would rave about D'Ambrosia, calling him "a fine young man" and boasting about how they were all going to make a lot of money. He put him in touch with Pete the Crumb Caprio up in Newark after Caprio said he had some Genovese people who were interested in the video poker machine business in Philadelphia. Positioning D'Ambrosia as an underworld financier, Natale would massage his ego whenever they were together.

Privately, however, he referred to D'Ambrosia as "Plugs," a reference to a less than successful hair transplant the young bookmaker had undergone. It was one of nearly a dozen nicknames Natale coined to describe his associates. Most were snide, mean-spirited comments made when the associate wasn't around. Those who were in on the conversation were expected to laugh. D'Ambrosia was "Plugs." Borgesi was "Freckles." Alphonse Parisee, an old-time gambler from Northeast Philly, was "Pipes and Slippers" because he liked to stay at home. Angelina, a burly wiseguy who liked Chinese food, was "Noodles" or "the Penguin." Ray Rubeo, whose hair was thick and kinky, was "Brillo Head." Previte was one of the few associates whom Natale did not belittle. He referred to him as "Mighty Joe Young."

Previte and D'Ambrosia shared a common interest in the gambling business, and as they spoke one day in October 1997, they realized they also shared a disdain for Skinny Joey Merlino.

"Joey's gonna be a shyster forever and ever," Previte said.

"He loves to gamble, that's his whole fuckin' thing," D'Ambrosia replied.

"A fuckin' degenerate, that's gonna be his downfall," said Previte.

Previte talked about how Merlino was continuing to "bet into guys" despite being reprimanded by Natale a year earlier over the $212,000 he ran up on Previte's book. Even though Previte was doing business with the FBI's money, there was a part of him that was still upset with Merlino's arrogance. Previte did not like to look the fool. And while he knew he would eventually have the last word, in certain underworld circles there were mobsters who believed Skinny Joey had gotten over on Big Ron.

D'Ambrosia understood what that was like. He said he wasn't sure, but he thought he might have taken a few hits from Merlino, who often bet through other gamblers or with a phony name. One week, he told Previte, one of his workers had taken "ten dime reverses" from the same gambler. A dime is a one-thousand-dollar bet. A reverse is a combination wager involving two games. If you pick both winners, you double your earnings. If both your teams lose, your losses double. A split, and you lose the edge money to the bookmaker.

"Nobody, not even Donald Trump, would bet ten reverses on the same day," Previte said. The only person who would make that kind of wager, he said, was somebody who didn't expect to pay if he lost.

"Why's he even doing it, when he can get a piece the other way?" D'Ambrosia said. "That's what I don't understand."

It wasn't about running a business, Previte said, it was about fast and easy money. "Joey don't want to grind nothing out." Previte complained that Merlino still had not admitted that he owed him the $212,000. D'Ambrosia said he also heard that Joey had bet into a book run by Steve Mazzone and Johnny Chang, two of his close associates.

"He's so fuckin' greedy," D'Ambrosia said.

Betting into somebody, or "guzzling," was a scam that all legitimate bookmakers were constantly watching out for. There were always con men and shysters trying to earn a living by guzzling a bookie, but word got around fairly quickly,

and eventually nobody would take their action. It was an entirely different matter, however, when the underboss of the
mob was routinely betting into whoever would take his play,
even people in his own organization.

"I'm only going to shoot into a guy who's got it coming," said D'Ambrosia. "Like he owes one of my guys
money, indirect. Or he's a rat. That's the only time I'll
shoot into a guy."

Not only was Merlino "shooting into" bookmakers, he
was also running his own gambling operation into the
ground. Merlino and Ralph "Ralphie Head" Abbruzzi had
come to D'Ambrosia asking to borrow eighteen thousand
dollars they needed to cover their losses after a week when
their customers picked a slew of winners in college and pro
football. "They went out of business," D'Ambrosia said of
the book Merlino and Ralphie Head were trying to run.
"They got killed."

"They don't pay nobody," Previte said. "They're not legit.
Nobody's gonna fool with them."

"They don't pay the pope," said D'Ambrosia.

Bookmaking is a long-term business proposition, he said.
You had to put money aside each week when you won to
prepare for the occasional week when you might lose. No
matter how good you were at setting the line, there would always be that week when something crazy happened, when
all the gamblers suddenly became Jimmy the Greek.

Previte nodded his head in agreement, adding that Merlino and the shysters he had around him didn't understand
any of that. "They make a steal one day, they go blow it," he
said. "They wanna take a pay every fuckin' day. You should
only draw a pay at the end of the week." And even at that, he
said, you had to put a big piece of it away.

"Eighty to ninety percent," said Danny Dee, who estimated that a successful bookmaker had a cushion of "a
quarter million dollars" set aside before he started "drawing
a pay."

"You just keep that there for the business," he said. "If you want to build a nice business."

Years ago, in testimony before the New Jersey State Commission of Investigation, mob informant Tommy DelGiorno, a one-time capo in the Scarfo organization, had tried to explain to authorities how important gambling was to an organized crime family.

"It's the wheel around which everything else turns," DelGiorno had said.

A smart bookie who knows how to set a line is going to make money, provided he's willing to work at it. A mob bookie is also going to use the gambling business as a jumping-off point for other things. In corporate America they call it networking.

Gamblers are degenerates. Most lose. When they lose, they need to borrow money. This borrowed money they use to pay their living expenses—replacing the money for food, clothing, and shelter that they've gambled away. They also use it to bet some more. The borrowed money usually comes from a mob loan shark at a killer rate. That puts the gambler in an even deeper hole, so he tries to negotiate his way out through a kind of barter system. He'll come with a proposition in which both he and his creditors can make some money. Maybe he's got a line on some hot stuff; maybe he knows about a business executive who keeps lots of cash stored in his house; maybe he can set up a sweetheart deal on a construction project for a mob-linked company.

Every day, Ron Previte got those kinds of propositions from people he knew. This was life in the underworld. Nine out of ten went nowhere. But the one in ten that resulted in a score was worth the effort. It was the way the system worked. And it all started with gambling.

Joey Merlino never grasped the concept. He just wanted the money.

12

Michael Casolaro was one of the best and busiest book-makers in the city. He had a stable of well-heeled customers, some from as far away as Las Vegas. He took tens of thousands of dollars in action on a busy sports weekend. During the football season, friends knew never to call him between noon and 4 P.M. on Saturday and Sunday. He was working. During the NCAA basketball tournament he was unreachable.

Thin and soft-spoken—he looked and sounded like the actor Aidan Quinn—Casolaro had been taking bets for more than twenty years when Joey's people came knocking on his door. He had started out writing numbers, then moved up to the more lucrative sports betting. He came by the trade honestly. Both his father and his grandfather had been book-makers in South Philadelphia. To them, it was a profession. There were shoemakers and plumbers, bricklayers and teachers. And there were gamblers and bookies. No big deal. They were all part of the neighborhood.

"It don't make you a bad person 'cause you're a gambler," he later told a federal jury.

When he was asked on the witness stand how his operation compared with others, he shook his head. "I don't ask about other people's businesses," he said. It was a classic street-corner reply. Nobody needs to know. A raid on his operation by federal authorities, however, offered a snapshot. During one five-month period, two phones registered at a Casolaro bookmaking location had logged forty thousand calls. A "work-sheet" seized by the feds included "accounts" ranging from \$5,780 to \$132,460. Those figures, Casolaro said later, represented either money he owed to a customer or money a customer owed to him. Sometimes, with a particularly good customer, he might wait until the end of a particular season to settle up.

Casolaro made a good living, but he worked at it. Like D'Ambrosia, he was serious about what he did and honest about his business. "I never robbed anybody in my life," he said in frustration and anger as he explained his encounter— he would call it a "nightmare"—with the Merlino mob.

Casolaro began paying the Merlino organization in 1993. It started with a fifteen-thousand-dollar "loan" to Joey, who said he wanted to start his own book. Casolaro came up with the cash, even though he knew he wouldn't get it back. He also made regular Christmas package deliveries. He said he could live with that. He had paid Scarfo. He had paid Stanfa. Now he would pay Merlino. Highly visible bookmakers, guys who were well established and had a lot of time and money invested in their businesses, knew they had to deal with the mob. Casolaro was a realist. "If you want to play, pay," he said.

He was on Merlino's list even before the big sit-down where Natale and Skinny Joey reestablished the street tax throughout the underworld. Each December Casolaro delivered an envelope with five thousand dollars in cash to Marty Angelina, one of Merlino's closest and most loyal followers. He also edged off some of his work to a mob-controlled sports betting ring run by a guy named Louie Sheep.

But in the summer of 1997, the wiseguys upped the ante.

Casolaro was invited to a meeting at John Ciancaglini's house. Casolaro and Ciancaglini went back years. Their fathers and grandfathers had been friends.

Ciancaglini, Louie Sheep, and Steve Mazzone were waiting when Casolaro arrived. They had a proposition for him. They wanted to "go partners" with Casolaro in his bookmaking business.

"I didn't want to do it," Casolaro later told authorities. "I told him [Ciancaglini] it was a bad idea."

The mob thought otherwise. They set up a "fifty percent" book. Under this arrangement, Casolaro would do most of the work and carry most of the financial load. In addition to his regular customers, he would handle action sent his way by Ciancaglini, Borgesi, Sheep, Mazzone, and several other mobsters. He was responsible for paying the gamblers who won. In effect, he was the bank. He and the mob would split the profits from the gamblers who lost. Casolaro's only safeguard was the so-called red figure. If the operation was losing money, there would be no profits to split until the debt—the red figure—was recovered.

He didn't like the deal.

"I knew John for a long time," Casolaro said. "I told him I really didn't want to be involved in the situation and in the operation like that because I didn't think I really had any chance of coming out ahead anywhere in the game."

Casolaro said he knew that "mob guys have a reputation of not paying . . . [that they] say, you know, 'Forget about it.'" Casolaro said he based his business on "dealing with gentlemen and professionals and people who could afford to bet." As a bookmaker, he said, he believed it was his responsibility to keep his customers "under control," to keep them from betting over their heads.

Under the scenario being set up by the mob, Casolaro knew he would have no control because he wouldn't know whose bets he was taking. That made him an easy target for the classic Merlino scam, which Natale would so accurately describe later from the witness stand.

"Joey had a great way," he said, shaking his head and smiling. "He'd find a bookmaker. He'd send a lot of bets through somebody else. He'd bet, bet. If he won, collect the money. If he didn't, he'd say he's not gettin' paid. And the guy couldn't make a beef or else Joey would take half his business. That was his game. He did it to a couple of his friends, also."

In the summer of 1997, Michael Casolaro saw it coming, and he still couldn't do anything about it. Johnny Chang said he couldn't help him. Casolaro's options were these: take on the new partners, or get out of the bookmaking business.

That first year, from the football season through the basketball season, Casolaro's red figure reached two hundred and fifty thousand dollars. He was able to work some of the debt down: gamblers run hot and cold, and during a cold streak the debt—on paper at least—was reduced. But the operation still wasn't making any money.

Eventually he learned that both Borgesi and Sheep had been putting in bets through third parties, collecting only if they won. If they lost, the money got carried on the books.

They couldn't lose.

It was no way to run a business.

In the old days, Angelo Bruno—who had come up as a bookmaker and loan shark—had a strict policy within his organization prohibiting members from betting with bookmakers affiliated with the mob. Bruno knew how important gambling was to the entire operation of a mob family. Like DelGiorno, he believed it was the wheel around which everything else turned. Money generated from gambling and related loan-sharking fueled the underworld economy, providing the financing for both legitimate and illegitimate operations that in turn brought in ever more cash.

But the gambling business only operates if those who win can collect and, more important, if those who lose pay up.

Gamblers only think about winning, Casolaro said in explaining the daily dilemma he faced in running a legitimate operation. A real bookmaker, on the other hand, worries first about how much he might lose. And if he wins, then he worries about whether he's going to get paid. It's a constant balancing act, one that is thrown completely out of whack by bettors who are merely taking a shot, bettors who have no intention of paying if they lose. The problem is compounded because those bettors usually risk more money and take more chances, betting long shots and ridiculous combinations.

Casolaro said he dealt with the nightmare for two years, slowly "weeding out" the phony bettors brought to him by Borgesi and Louie Sheep. Still, the operation never turned a profit. And on top of that, he said, he was still required to deliver a Christmas package. In fact, in 1998 Angelina demanded ten thousand dollars rather than the five thousand dollars Casolaro had been paying. Casolaro argued that the mob knew his business was running in the red and ought to cut him a break. Ciancaglini said that wasn't possible.

"I thought there should have been a little fairness and team play," said Casolaro, who asked for a meeting with Merlino.

Angelina set it up at the Philadium, a bar in the 1600 block of Packer Avenue, down near the stadium complex in South Philadelphia. Merlino attended with several other members of the organization. Casolaro brought his betting sheets, hoping to use the figures to support his plea.

"Can I talk to you for a minute?" he asked Merlino.

"No," Skinny Joey replied.

Casolaro ignored the rebuke and tried to make his case. He said he was borrowing money to stay in business and would have to borrow more to meet the Christmas shakedown payment. But Merlino didn't want to hear about it.

"There were a few guys to my left standing at the bar that I knew were with him," Casolaro said. "I didn't want to pursue the issue, since you never knew when a flying bottle . . . I reached in my pocket to show him the sheets, but it never got that far."

Merlino looked Casolaro in the eye. "No," he said. "And this is the last time I'm gonna tell ya."

Casolaro paid the ten thousand dollars that Christmas.

That spring he got into a confrontation with Angelina, who came to his house and threatened him with a baseball bat. It was all a misunderstanding, but Casolaro steered clear of Angelina for the next several months, refusing to show up at a meeting at a local bar where, he later learned, Angelina and about a dozen others were waiting for him.

"They expected me to go down there with a bunch of guys," Casolaro said as he recounted the incident several years later. But he never even thought about showing up for the meeting. "I'm in my forties. That's all kid stuff, like guys wanting to fight other guys. . . . There were like fifteen guys there ready to go. . . . It's all a big impression thing, you know, who's tough, who isn't. . . . You come to my house with a baseball bat, I really got nothing to talk to you about. If I don't see you again, that would be fine for the rest of my life."

Michael Casolaro made one more Christmas payment. This one, in 1999, was for five thousand dollars. He gave the envelope to Ciancaglini, asking him to pass it on to Angelina. By that point, Casolaro had the feds on his back.

The FBI had raided his bookmaking operation early in December. He later said he knew they would be coming from the moment he was forced to go partners with the mob. "They didn't come for me," Casolaro said. "They came because of my affiliation."

Previte could appreciate what both D'Ambrosia and Casolaro were going through, the snares and snags they had to maneuver around to stay in business. At the time, though, his own bookmaking operation was being fronted by the FBI; wins and losses were no longer a major concern. Beginning in 1997, the feds were manning many of the phones that took the action sent Previte's way by the Merlino-Natale organization. It was part of an ever-broadening federal investigation.

Previte's book operated out of the Hammonton area in South Jersey. But the bettors that Merlino and Natale were providing didn't want to make long-distance calls. They wanted Philadelphia numbers. With the FBI's approval,

Previte got a small apartment in South Philly where he set up shop. The feds arranged to have the phones installed. Sometimes Previte would have workers there taking the calls. On other occasions he would arrange for any calls coming into that location to bounce over to his Hammonton headquarters.

"I had a guy, Johnny Beebe, used to work for me," Previte said. "He was in the Philadelphia spot a couple of times. He was there when the new phone lines were installed. But he looks and he sees that the phone company order sheet says the phones are being installed for a federal agency. He gets nervous and comes to me. I hadda make up some story. I told him Ralph used to have connections with the government, and that's what we used to get the phones. I couldn't fuckin' believe it.

"The agent who set up the phones apologized. He said he didn't realize. He wasn't thinkin'. I'm out there with my ass on the line and he's not thinking. That's the kind of shit I hadda put up with. I always had to watch out for myself. If I knew then what I know now, I'm not sure I would have ever made the deal with the FBI. In a lotta ways, they're smart. But in other ways, they're dumb.

"Beebe never brought it up again. But he would never work out of that apartment either. He didn't want to use those phones."

Although in his forties, Johnny Beebe was from the old school, a throwback to a time when people who operated in the underworld did all they could to avoid having their voices crop up on tape. Carlo Gambino, the legendary New York Mafia don, shunned using phones for most of his life because of his concern over wiretaps. Angelo Bruno was just as circumspect. Gotti, on the other hand, buried himself with his own words, picked up by electronic listening devices hidden in the Mulberry Street apartment above his clubhouse. Stanfa is serving a life sentence in large part because of conversations he had in his lawyer's office. The Racketeering Influenced and Corrupt Organization (RICO) prosecutions that have decimated the American Mafia over the past twenty years have been built around two key components: the turncoat testimony of former mob members and recorded conversations picked up on phone taps, room bugs, and body wires. Both were in play as Previte helped the FBI build its case against Merlino and Natale.

But prosecutions were just part of the problem an incriminating tape could cause. Another danger in having your unguarded conversations recorded and played in open court was underscored early in 1998 when Anthony Turra went on trial in federal court in Philadelphia. Tapes, it would turn out, could be deadly.

The uneasy peace that Natale had established between the Turras and Merlino in the summer of 1995 held up until the Turra trial, a case built around a racketeering and drug-trafficking indictment that had been handed up a year earlier. The same FBI informant who had been feeding the feds information about the plots to kill Skinny Joey back in 1995 had also been giving up information about the Turra drug operation.

In 1997 the Turras—Louie, Rocco, and Anthony—and most of their top associates were named in a sweeping federal indictment that charged Louie Turra with heading a criminal enterprise that trafficked in narcotics and that had conspired to kill three individuals. One of those was Skinny Joey Merlino. By the time the trial started in February 1998, half a dozen defendants had pleaded guilty. Of the Turras, only Anthony was still a defendant.

Rocco Turra, in a move that shocked both law enforcement and underworld figures, had turned government witness and was ready to testify against both his brother and his nephew. Louie Turra, looking at a possible life sentence if convicted, chose another way out. He killed himself. On January 7, just weeks before his trial was to start, the handsome young drug lord—a kingpin who authorities alleged had made millions dealing coke—wrapped a bed sheet around his neck and hung himself in his federal prison cell at the Metropolitan Correctional Center in New York. He was thirty-three years old.

Anthony Turra, sixty-one, was one of five defendants who went on trial when the case began. He was facing one count of conspiracy to murder Joey Merlino. His lawyer thought he had a good shot to beat the rap. Because of his medical condition, Turra was the only defendant out on bail. Looking forlorn, walking with the aid of a cane, he shuffled into court each day to join his codefendants at the defense table. Among other things, he suffered from terminal lymphatic cancer—he had undergone several rounds of chemotherapy—congestive heart failure, kidney disease, and advanced emphysema. In a daily ritual that was an act of either defiance or resignation, he slipped outside during a break in the trial to smoke a cigarette.

The government's evidence included the highly incriminating testimony of both Rocco Turra and the informant who had been cooperating since 1995.

The feds also had tapes. Lots of tapes. And on several the jury heard about the plots to kill Merlino—heard the anger and the venom and the disdain coming from the mouths of the late Louie Turra and his father, Anthony. "This one guy's gotta go . . . Joey Merlino," Louie Turra told William Blackwell, the government informant, during a conversation Blackwell recorded shortly after Turra had been savagely beaten at Libations, the after-hours club, by Merlino's associates.

"I've been thinking of banging them right at their fuckin' hangout," Turra said, itching to kill Merlino and his friends at the Avenue Café. Turra wanted to crash into the café on a motorcycle, pull out an Uzi, and open fire. Blackwell, from the witness stand, said the younger Turra never got over the beating he had taken—that he constantly talked about getting even. During another taped conversation, Turra drove Blackwell around South Philadelphia and showed him where Merlino and his girlfriend, Deborah, were then living.

But the most damaging tape, from an underworld perspective, was a conversation Anthony Turra had with his son and Blackwell. Anthony Turra suggested that hand grenades be thrown through the window of Merlino's home.

"Dump two or three grenades in the fuckin' house," he said.

When Blackwell interrupted, pointing out that Merlino's girlfriend lived with him, Anthony Turra said it didn't matter.

"Fuck his girlfriend. . . . She's a scumbag enough to live with him. Then she don't deserve living either."

The tapes and testimony made headlines in the local papers and topped the nightly news reports. Merlino heard it all and wasn't happy. "He told me we have to kill Tony Turra," Natale testified later. "He went off on a torrent." Natale said he realized Merlino was right, that something had to be done. The fact that he had grown up around the corner from Anthony Turra no longer mattered. This time, personal loyalties and friendships carried no weight. This was an insult to La Cosa Nostra.

"It's one thing to go after a man from La Cosa Nostra for yourself," Natale explained. "But when you hurt someone in his family, that's an infamy. . . . He wanted to throw a bomb into the house and hurt Deborah, or maybe the child if she was pregnant. That couldn't be. . . . It could never be."

Natale said he okayed the hit. "If we allowed that, then anyone in the city of Philadelphia could do it to anybody in La Cosa Nostra. So we had to make an example, right then and there."

A few days later, in a bar in South Philadelphia, Natale said he met with Merlino, Borgesi, and Mazzone and gave his final okay. Meeting in a back room, he said he told them, "Take care of this. . . . Let's get it done now. We'll set an example. . . . Then no one could ever hurt anybody's family or talk about hurting anybody's family."

As with all the other murders and attempted murders he testified about, Natale had little to back up his version of what was said. As for the details of the hit itself, he had to rely almost entirely on what he had been told. "You pick out the shooter. You do what you gotta do," he said he told Merlino at that meeting.

It was a reprise of the Billy Veasey and Tony Milicia hits. Natale said he got a heads-up the day before the murder from Borgesi, who said everything was ready. "It's all set for tomorrow morning," he said Borgesi told him. "We're going to do Tony Turra."

"Good," Natale replied. "Be careful."

Jury deliberation in the racketeering trial had begun on March 17, a Tuesday. The jurors were due back in court to resume their discussions at 9:30 A.M. on March 18. Anthony Turra, as he had throughout the trial, left his house that morning around 8:30 A.M. Using his cane, he gingerly made his way down the four stone steps that led from the front door of his two-story brick row house near the corner of Twentieth and Wolf Streets in South Philadelphia. Just as he turned to head up the sidewalk, a burly man in a ski mask walked up, pulled out a gun, and opened fire. Turra was shot twice, once through the eye and once in the back. The shooter ran to the corner, jumped into a waiting car, and disappeared.

Anthony Turra was rushed to the Hospital of the University of Pennsylvania, where fifteen minutes later he was declared dead. Among the things recovered from Turra's pockets were a prescription bottle containing thirteen one-milligram tablets of lorazepam, a drug commonly prescribed to ease anxiety, and a box of Marlboro Lights.

In less than three months, Louie Turra had killed himself, Rocco Turra had taken the stand as a government witness, and Anthony Turra had been gunned down on the sidewalk in front of his home. A unique South Philadelphia crime family was no more.

According to Natale, he met Merlino and half a dozen other associates at a South Philadelphia row house a few days later to celebrate. "Everything went like clockwork," Merlino told him.

Just as in the Billy Veasey hit, there were lookouts with walkie-talkies and "blockers" standing by in cars to cut off exit routes and ensure a clean getaway. Natale said he was told the shooter was a young mob associate, Michael "Mikey Penknife" Virgilio.

When he met Virgilio, Natale said he hugged and kissed the young wiseguy, congratulating him for a job well done. Virgilio was called "Penknife" in a street-corner homage to his father, the late Nicholas "Nick the Blade" Virgilio, a Scarfo crime family soldier who had died in prison. Asked about his nickname while on trial in a racketeering case in the 1980s, Nick Virgilio had smiled. He was called "the Blade," he said, because "I was always a sharp dresser."

Yet despite Natale's account, Mikey Penknife Virgilio, who occasionally stutters and who takes medication for Tourette's syndrome, was not charged with the Turra murder in the racketeering case in which Merlino and the others were later indicted. Virgilio has denied any involvement in the shooting.

Four days after the murder, Skinny Joey and most of the mob took part in yet another public celebration. On March 28, at a restaurant/catering hall called Colleen's, Merlino and Deborah threw a party to commemorate the baptism of their second daughter. Not quite as lavish as the celebration at the Ben Franklin House, the affair was attended by about two hundred and fifty close friends and associates. Outside, police and FBI surveillance cameras got it all. Sophia Merlino had been born on January 6, 1998, the day before Louie Turra committed suicide.

As the summer of 1998 approached, Ralph Natale was preparing to reap the benefits of his latest scheme. By his own estimation, he had funneled about fifty thousand dollars to Milton Milan, the new mayor of Camden, New Jersey. Most of the money had been paid in cash, large bills stuffed in envelopes and delivered by Danny Daidone.

Daidone had set up a construction consulting company in the old Sears Building on the Admiral Wilson Boulevard, a five-minute drive from City Hall, and had been part of a group that backed Milan's run for the city's top office. A Marine veteran of Puerto Rican descent, Milan was the city's first Hispanic mayor. Natale had begun to woo him a year earlier, when Milan was president of the city council. Throughout the mayoral campaign, Natale, through Daidone, was a silent political backer.

Camden was in line for millions of dollars in state and federal aid. Whoever controlled City Hall would control the flow of that cash. One of the most depressed cities on the East Coast, Camden is home to about eighty thousand residents. Predominantly black and Hispanic, they have watched despondently as one administration after the other—white, black, and now Puerto Rican—squandered opportunity after opportunity to make the city work. Carrying on a legacy of incompetence and greed, Milan would be the third mayor in the last twenty years to be convicted of federal corruption charges. But in 1998, he and Natale appeared to be sitting on top of their respective worlds.

Natale told Daidone to tell Milan, "If we do well, he'll do well." By that point, the feds were tracking every move Natale made in Camden. They had dozens of tapes from a host of bugs and phone wiretaps. They had surveillance videos. And they had a paper trail. Among other things, they knew that in January 1998 Daidone had paid a travel agency $1,433 to cover a vacation trip to West Palm Beach for Milan and his wife. Daidone and his girlfriend accompanied the newly elected mayor.

The quid pro quo had not yet kicked in, of course. Natale kept waiting for one of his surrogate companies to be

awarded a city contract. He had two minority front companies lined up. One, a "double minority," was owned by a
Hispanic woman. Natale had Daidone running around behind the scenes "expediting" the process. On tape he referred to Daidone as his "chief of staff."

Daidone had scored a memento that would come back to
haunt Milan: an autographed copy of the slick program dis-
tributed at Milan's July 1, 1997, inauguration party.
Scrawled across a picture of Milan, felt-tipped block lettering read "To my main man Dan Daidone. Thanks for everything. Milton Milan." In that same brochure was a full-page
ad taken out by Wholesale Seafood, Inc., of Gloucester
City. Daidone was the company's secretary-treasurer, Natale a "salesman." The mob had not yet busted the company
out. The day after the inauguration, Natale picked up a
nine-hundred-dollar dinner tab at the Italian Bistro, a
restaurant in Cherry Hill where Milan and some other
politicos, including Daidone, had gone to break bread and
plot strategy.

Natale had stopped by the table, shook the mayor's hand,
and congratulated him. The best, he thought, was yet to
come. The feds had other plans.

"One of the problems," Previte said, "was that as long as
Ralph was on the street, I couldn't get close to Joey. At first
this was good. It was protection for me. In the beginning, I
think they woulda killed me.

"But now we're all making money. I'm giving 'em a lot
of cash, but I'm not directly involved in too many things
Joey's doing. I got a lot of good tapes on Ralph. The meth
case is a slam dunk. But with Joey it was mostly some gambling talk and swag."

Previte had touted himself as a fence, claiming he could
move stuff that was coming off tractor-trailers. In 1997 he
had come up with seven hundred and fifty dollars and taken
twenty-five of the ceiling fans Merlino's guys were moving
from one of the first tractor-trailer heists. Eventually he and

Joey would set up a couple of bigger deals. But first Ralph had to be removed from the picture.

In June, Natale was "violated"—cited for meeting with convicted felons while on parole. Among other things, the feds disclosed for the first time that the party at Gianna's in March 1997 had been bugged and surveilled. That alone was enough to send Ralph packing. He was taken to the federal prison in Fairton, a South Jersey facility about sixty miles from Philadelphia. From there he would be transported to another prison in Ohio.

Speculation was that he would do a year to eighteen months for the parole charge. But there was more. Rumors spread through law enforcement and underworld circles that Natale, then sixty-one, was never coming home. The feds were sitting on a big drug case. As a twice-convicted narcotics trafficker, if Natale were convicted a third time he would get life.

With Ralph off the streets, Merlino finally assumed the role of acting boss. Mazzone was bumped up from consigliere to underboss. Borgesi moved from capo to consig. The guys from Twelfth and Wolf were now running the mob.

13

The seduction was built around the one thing Skinny Joey Merlino couldn't resist—money.

Even before Natale had been violated, Previte, who knew what was coming, had planted the seeds.

Early in June, each wiseguy who had been at the Gianna's affair—Natale, Merlino, Borgesi, Previte, and about a dozen others—had received a letter in the mail or hand-delivered by an FBI agent, notifying them that they had been intercepted by the court-authorized electronic listening device during the party.

This created a buzz throughout the underworld. Merlino, who had talked to three or four lawyers, saw it as a good sign. If they were going to be indicted, he said, they would not be told about the intercepts until after they were arrested. The fact that they had been notified meant, in his opinion, that the investigation had turned up nothing of substance. By law, he said, the feds had to notify anyone picked up on a wire once the probe was completed.

"Every lawyer I talked to said they think it's good," he told Previte as they stood talking on a South Philadelphia street corner on June 10, 1998, two days before Natale was arrested for violating his parole.

"I heard the whole joint was wired," Merlino said of the restaurant.

Previte, who had provided the feds with the information

they needed to get the court authorization to bug Gianna's, played along.

"Yeah, so that means somebody said something ahead of time."

"Right," said Merlino. "And they put a television outside. We're intercepted."

"You know what this means for the other guy, don't ya?" Previte said, referring to Natale.

Merlino knew what Previte meant—that Natale could be pinched for violating his parole—but he said he didn't think that was going to happen. "They could have violated him five years ago if they wanted to. They don't want to violate him. . . . They got shit bugged. If he's in jail they ain't gonna hear nothin'."

Cars were bugged, apartments were bugged, phones were bugged, Merlino said. The feds wanted Natale out on the street talking. Merlino said he was disgusted with it all. "Motherfuckers." He told Previte he had met with Natale the day before and complained.

"I told him forget about these parties, forget about sittin' around. There's no need for it. You wanna get fuckin' pinched? You see they're buggin' everything. Why talk about . . . Fuck, we don't need this.

"If I gotta see you in case of an emergency," Merlino said he told Natale, "I see you." But unless it was a matter of "life or death," Merlino said, no more meetings.

Previte, picking up on an opening, nodded. "Every time I'm around him there's trouble," he said.

Everyone in the underworld had been gossiping for weeks about a big blowup between Natale and Viesti down at the Fireside Restaurant in South Jersey. The source of the dispute was Ruthann, Ralph's young girlfriend. Previte was there when it happened.

"Ralph and Weasel got into some argument, and in the middle of it Ralph brings up Ruthann," Previte said. "He says he doesn't like the way Weasel has been acting toward her. Then he says she's a princess. The same shit as always. I'm looking down, trying not to laugh.

"All of a sudden, Ralph stands up and slaps Weasel in the face."

Previte described the incident in several taped conversations, including one in which he told two other wiseguys, "All bitches are treacherous, but she's no fuckin' good. . . . I seen her get that stupid Weasel slapped in the face. Although he needs a slap, but you don't slap a guy because of some bitch. . . . You understand? I don't give a fuck what a piece of shit he is. You got him around you. And . . . you slap him to impress her? That's bullshit."

Now Previte picked up on the same theme with Merlino. "Fuckin' guy won't let me do nothing, Joey. . . . I can't make ten fuckin' cents."

"That broad's got him nuts," Merlino replied. ". . . He's in love with her. I don't know what the fuck it is."

"In love with what?"

"I have no idea. . . . No idea . . . It's a shame too, really. But his head ain't right now. I mean, his head ain't sharp."

"Naw, he ain't sharp," Previte said.

"The fuckin' broad," said Merlino.

Merlino said he thought it was a bug on Ruthann's phone that led the feds to the party at Gianna's. "Why does that girl got to know that we're having a party?" he asked. "I think it was on her phone that they did the fuckin' wire."

"I wouldn't doubt it," said Previte with a straight face.

Previte decided to up the ante in the conversation. "He's history," he said of Natale. "I mean, don't say I said that, but he's fucked up. I mean . . . the guy always went to bat for me. I can't say nothing bad about him. I love him, but . . . he's a fuckin' mess."

"It's bad," said Merlino.

"Can't make no money," said Previte, getting to the heart of the matter.

Previte said he had three or four deals on the back burner,

waiting for Natale to approve them. Shakedowns, extortions, the sale of hot stuff. He said he had a guy who could get him all kinds of jewelry. He then asked Merlino if he had anything for him.

Merlino said he was waiting to hear about a stolen shipment of vodka, a "couple hundred cases." A contact in Rhode Island, he said, had a line on the stuff. But Merlino said he didn't know about the quality. If it was Stoli or Absolut, he said, they could make some money on the deal.

"I'll take it all," Previte said. "Just as long as you give me a good price. . . . I can blow any hot shit out. Believe me, I can blow it out."

Merlino liked what he was hearing. And with or without Natale's knowledge and approval, he said they'd go forward. It was just what Previte and the FBI wanted him to say.

"We gotta start makin' money," Merlino said. "I gotta start makin' money."

"I always did whatever I was told," Previte said. "This guy ain't got me doin' nothin'. That's why I'm comin' to you."

"Go do what the fuck you want to do. He ain't gonna even know. If he knows, he finds out, I'll tell him I told ya, I gave ya the okay. . . . Forget him. . . . You go do what the fuck you want to do. . . . You got my word . . . on my kids. I'll just tell him that I gave you the okay. Go do what the fuck you want to do. Make some fuckin' money."

Two days later, Natale was off the streets.

U.S. marshals picked him up around 11 A.M. on Friday, June 12, just as he was coming out of Cooper River Plaza South, his apartment complex in Pennsauken. He was charged with associating with known felons. The parole violation document cited twelve different instances over the past year and a half, including meetings at the Palm, the Garden State Racetrack, and the party at Gianna's.

Less than a week later, New Jersey authorities announced the indictment of Robert Constantine on drug and weapons

charges. In addition to being charged with manufacturing
and distributing meth, Constantine, who had a prior drug
conviction, was cited for possession of a .45-caliber handgun by a convicted felon. The gun was found during a state
police raid at his home in Sea Isle City, where investigators
also found a handwritten formula for the manufacturing of
meth. Investigators also raided a small storage garage Constantine had rented in nearby Dennis Township, where they
found over a gallon of P2P and six ounces of meth. The indictment, authorities in New Jersey said, was part of an ongoing investigation. The press release detailing the case did
not identify Constantine as Natale's son-in-law, but everyone knew the connection.

What Merlino and others also knew was that Natale had
been using his son-in-law to cook meth. The question was
no longer if, but when, Natale would be facing drug charges
of his own.

That summer, Previte began dealing with Merlino on a regular basis, sometimes meeting with him two or three times a
week. Attending parties. Hanging out in bars in South
Philadelphia and at the Jersey shore.

Although he didn't realize it at the time, Merlino was living his last summer of freedom. As was his custom, he and
his wife rented a condo in Margate, an upscale residential
community on the shore just south of Atlantic City.

There were about a dozen bars and restaurants in Margate, ranging from chic bistros to shot-and-beer joints
that looked like college fraternity basements. Almost all
the establishments catered to the twenty- and thirtysomething crowd, a mix of college students, young school-
teachers spending the summer at the shore, and junior
executives and career women sharing group rentals on the
weekend.

A city of about nine thousand year-round residents, Margate would swell to four times that size during the summer

months. Merlino and his entourage would spend weekends there, partying from Friday night until Monday morning.

Jerry Blavat, a renowned Philadelphia disc jockey, once described by investigators as "court jester to the mob," had a club called Memories in Margate, where Merlino and a dozen other young mobsters would hang out. Blavat, who billed himself as "The Geator with the Heater," had been a friend of mob bosses for more than twenty years. His mother and Angelo Bruno's wife had been close neighborhood friends. Both came from the same section of southern Italy. They were *comari*. After Bruno died, Blavat was tight with Scarfo. While not particularly close to Stanfa—the Sicilian-born mob boss could never understand the Geator's rock 'n' roll patter, or why he should be taken seriously—Blavat resurfaced around the wiseguys when Natale and Merlino took over, spinning records at Merlino's big coming-out party at the Ben Franklin.

College kids and yuppies from Philadelphia, the Pennsylvania suburbs, and New Jersey crowded Memories each weekend. But Merlino and his group were usually afforded the star treatment, given access to a bar in the back of the establishment that was off-limits to most customers. They could see and be seen, but they didn't have to mingle. For most of the weekend revelers, it was enough to know that they were in the same place as Skinny Joey.

Merlino, naturally, attracted the attention of local authorities. He would complain that the Margate police were out to "bust his balls," that he and those around him would be cited for any small infraction because of who they were.

That summer his wife, Deborah, and three women she was out partying with, including Ruthann Seccio, made headlines after the police found a marijuana joint in the sports utility vehicle they were driving. On another weekend Joey Merlino was cited for walking in public with an open container of beer. He had just exited a bar on Washington Avenue and, realizing he was drunk, decided to take a taxi back to his condo. Police spotted him and pulled the taxi over. He was given a citation for the open liquor container, and when he threw the citation on the ground he was given another one for littering.

Previte was amazed as he watched the young mob leader up close.

He knew Merlino was a "shyster," that he would rob and steal from almost anyone. He also knew that Merlino could be treacherous and, most important, that he was deadly.

But there was something about Merlino that was appealing. He could be funny and self-deprecating. He was a quick study. And when it came to people in trouble, especially kids, he had a genuine soft spot. Sure, the parties for the homeless were financed with other people's money. But Merlino's desire to do something for kids who otherwise might not have a Christmas was genuine.

"In a lot of ways, he had heart," Previte said.

Merlino had invited Previte to a party that summer in Ventnor. The affair was going to be held at the home of Merlino's aunt, who had a pool and a big backyard. They were going to celebrate the second birthday of his oldest daughter, Nicolette. Merlino told Previte to bring his girlfriend and her three kids. "I got a fuckin' pool. It's warm. . . . I got clowns, horses, every fuckin' thing."

Previte told Merlino that his girlfriend's son was autistic, that he could be difficult to control. Previte had seen other people become uncomfortable in the kid's presence.

"It's all right," Merlino said.

At the party, Merlino went out of his way to make the boy, then about twelve years old, feel at home.

"We get there and the kid goes up to him and says, 'Hey, Joey Merlino. You're the gangster. You're the gangster,'" Previte said. "Joey just laughed. Later, everybody was in the pool and the kid got it in his mind that he wanted to clean it, use that scooper and the other equipment. Right in the middle of the party. Joey goes, 'Okay, everybody out of the pool.' And he helped the kid. He and Joey cleaned the pool. You could see it was real. He had heart.

"Look, I knew he would rob me if he could. I knew he *had* robbed me, and I knew he had wanted to kill me. None of that had changed. But there were times when he was a likable guy."

That summer Previte made a couple of dozen tapes for the FBI. Most dealt with swag, the hot stuff that Merlino was hoping to get and that Previte said he could move. Previte mixed in pieces of the truth with stories based on where the FBI was trying to move the investigation.

He and Merlino talked about putting some loan-shark money on the street. Merlino said he would hook Previte up with John Ciancaglini, that they would run the operation together.

Slowly but surely, Ron Previte was moving closer to the inner circle of the organization. He and Ciancaglini had never gotten along; they had bad blood going back to 1993 and Previte's role in the extortion of Ciancaglini's bookmaker while he was in prison.

"I sat down with Johnny Chang and I told him, 'Ronnie's our friend,'" Merlino told Previte during a conversation at the end of June 1998. "He said, 'I understand.' . . . He wants to sit down with you and shake hands. . . . I told him, you know, 'Bygones be bygones. Ronnie's a mover.' I said, 'He's gonna get us some beans to put out.' I said, 'You and him put it out.' He said, 'Beautiful.'"

"Beans" was a reference to the cash Previte said he could come up with for the loan-sharking business. "Blueberries" might have been a better image: Previte had always used a group of very wealthy farmers in the Hammonton area as a source of cash. It was part of what Previte found so charming and endearing about his hometown. The Chamber of Commerce might not agree, but there was a subculture there that allowed Previte to flourish. It was part of that old-time Sicilian mentality, a lifestyle built around a blanket mistrust of authority, an unwavering sense of independence, and an unflagging belief in self-reliance.

"When I was running my book, I wouldn't always edge off to other bookmakers," Previte said. "I had these old-time farmers who would always hold some action. And after the growing season, when they were flush, I'd go to them to get money to put on the street."

It was better, the old-timers knew, than putting their money in a bank. The interest rate was higher Previte's way. And the government had no way of tracking their earnings.

Once the blueberry crop was in, Previte told Merlino, he'd be able to get his hands on the cash to underwrite the loan-sharking operation, get it off the ground and up and running. After that, it would be self-sustaining.

A good loan-sharking business is a guaranteed money-maker, as long as it's run properly. A guy borrows ten thousand dollars for ten weeks at, say, three points, meaning he's got to pay three hundred dollars a week in juice—interest for the length of the loan. At the end of the ten weeks you've collected three thousand dollars and the guy still owes you the ten grand principal. It's a beautiful business, provided you have the clout and the status to ensure that the debtor pays up every week.

Merlino had the clout. What he didn't have was the money, the beans.

Previte was going to get it for him. Bygones were bygones.

Throughout that summer and fall and into the winter, Previte usually brought cash whenever he went to see Merlino. He'd hand him an envelope with a thousand or fifteen hundred dollars in it. It was a tribute payment. He had been doing the same thing for months with Natale. He knew it was the language that both mob bosses spoke.

On one occasion, however, he brought a Rolex watch instead. It was a Submariner with a diamond bezel—the kind that Sean Connery wore in several of the James Bond movies. It was worth about six thousand dollars. It came from the FBI, part of a stash of seized and forfeited jewelry that the feds now had at their disposal. Previte, of course, said it came from a "source," a guy who often got his hands on hot stuff. Merlino loved it.

"I got a bunch more stuff comin'," Previte told Merlino. "You got anyplace in Philly we can move stuff?"

Merlino said he knew three or four jewelers who would take anything they had. After trying on the watch, he said, "It's beautiful."

Previte laughed. "And don't sell it," he said. "Wear it."

"You kiddin' me?" Merlino promised he would. Then he asked about gold Rolexes. Previte said he might be able to get some.

"I can move 'em," Merlino said.

The image of the mob boss hawking hot watches was perfect. This was La Cosa Nostra at the turn of the twenty-first century: Merlino, in a trench coat, standing on a street corner. "Hey, buddy, wanna buy a watch?" The Mafia Don as a two-bit hustler.

Angelo Bruno ran the family for twenty-one years. When he died he was a millionaire. He had real estate in Florida, hidden interests in a company that ran casino junkets to London, a stock portfolio that included thousands of shares of Resorts International, the company that opened the first casino in Atlantic City. Bruno had bought the stock when it was selling for two dollars a share. After the casino opened, it rose to about a hundred and fifty dollars, then split three-for-one and rose again. Bruno made a lot more on that deal than Merlino could ever hope to earn hawking hot Rolex watches.

More to the point, Bruno saw no need to wear a Rolex.

"A Rolex watch is a sign of affluence to certain mob guys," Previte later explained. "It's big and it's gaudy. It's something on your arm that says, 'I'm somebody and you ain't.' That's why certain wiseguys like 'em. You take some bank executive, some president of a corporation, he don't wear a Rolex. He wears Cartier.

"But mobsters love Rolex watches. When I was a kid I wore a phony Rolex till I could afford a real one. Now I got three.

"It's status. I was watching the Johnny Carson show one night and a guest asked him about the three-thousand-dollar suit he was wearing. It was the preacher Billy Graham. And he goes to Carson, 'You don't really *need* that three-thousand-dollar suit, do you?' And Carson says, 'No. I don't need it. But I look good in it.' That's the way mob guys feel about Rolex watches."

Bruno never needed to feel important. He knew who he was. Merlino, Natale, and those around them weren't as sure. They needed the accessories, the things that made them somebody.

Previte also claimed that "some Russians" were providing him with cell phones, models that couldn't be traced or tapped. He got one for Merlino. The FBI had it tapped from day one.

That summer Previte would also lease a Land Rover for Merlino. Again, the FBI financed the deal.

Like Natale, Merlino was now in the crosshairs of an electronic investigation. Every move he made was monitored in some fashion.

Unaware, he continued to live the life of an upwardly mobile, socially active Mafia prince. In one conversation he complained to Previte about the difficulty he was having getting tickets to a Spice Girls concert for "this girl" he knew who was hot to see them. "Whoever the fuck they are," said Merlino. (The young mob boss's own taste, oddly enough, ran more to the likes of Elton John.)

Previte also joked about Merlino's latest brush with the New Jersey Division of Gaming Enforcement. In June he had been spotted gambling at one of the casinos, even though he was on the state's exclusion list. Gaming regulators said they had a video of Merlino at the blackjack table, irrefutable proof that he had violated the ban.

Merlino denied he was there. The tape was a fake, he said. "Look at Forrest Gump."

"What's gonna happen?" Previte asked. "You gonna plead guilty or what?"

"It's all bullshit," Merlino insisted. "Fuckin' tapes are fixed."

"Is that what you're gonna tell them?" Previte asked, laughing. "You're fuckin' nuts."

Actually, Merlino said, he would probably have to plead guilty and pay a fine. But it had nothing to do with the charges.

"I had a broad with me," he explained. If the tape were played at a court hearing, the media would get a copy. It would probably show up on TV. And his wife would see that he wasn't hanging out with the guys on the night in question.

"I got the fuckin' broad next to me, you know what I mean?" Merlino said.

Previte laughed again. "Oh, shit," he said. "Then you might *have* to plead guilty."

Ralph Abbruzzi, known as "Ralphie Head," was the mob's point man with the ring of hijackers who routinely drove tractor-trailer loads of cargo out of the storage terminals along the Delaware River.

A sometime bookmaker and longtime mob associate, Ralphie was one of those guys always hanging on the fringe of the organization, looking for a score. A federal prosecutor would later describe him as "a lifelong petty criminal [he had four prior arrests, most for gambling] who has never engaged in substantial legitimate employment," a "mob associate who is drawn to organized crime and gangster life like a moth to flame."

To most of the people who knew him, however, Ralphie Head was just a guy from the neighborhood.

That summer Merlino put him together with Previte.

Abbruzzi, then forty-eight years old, had lived all his life in South Philadelphia. He got his nickname because of his large, pumpkin-shaped head. He was the Charlie Brown of wiseguys. And like the *Peanuts* character, he got lost in the details.

One of Previte's first big scores of swag was a shipment of Raleigh bicycles Abbruzzi steered his way. Previte took over five hundred bikes, agreeing to pay a hundred dollars apiece for them, although they were valued between two hundred and twenty-five and two hundred and fifty. Previte and several "associates"—members of the Pennsylvania State Police and FBI—off-loaded the bikes onto their own trucks with the help of Ralphie Head and two other mobsters. It was a classic sting.

"They got five-year warranties," Abbruzzi boasted in one conversation picked up on tape. Previte didn't know what to say. "This was hot stuff," he said later, still laughing about the conversation. "What the fuck was I gonna do with warranties?"

But that was Ralphie Head. "The bikes are gorgeous," he added. "Latest colors and all."

Previte also took a tractor-trailer load of baby formula—Enfamil—that Merlino and Abbruzzi said was worth "more than the bikes."

Merlino told Previte there were 2,436 cases of the baby formula. With two young daughters, Skinny Joey knew about the price of baby food.

"A six-pack is worth twelve dollars," the mob boss said. "These are cases."

The FBI, however, didn't want to come up with anything close to the amount of cash Merlino was looking for. Previte waited a week. Then he told Merlino and Abbruzzi that he couldn't move the product, that his "buyer" was afraid to take it, afraid to put it on the shelf.

Previte told Abbruzzi he had abandoned the trailer, still loaded with the baby formula, at the J. Fenimore Cooper rest stop on the New Jersey Turnpike. In a series of conversations he taped in late July 1998, he and Ralphie Head discussed the location of the trailer and the hapless attempts of the original hijackers to recover the load.

They couldn't find the truck.

Abbruzzi said the hijackers thought they were being stiffed by the mob, that he and Merlino had sold the baby formula and were pocketing all the cash.

"Ron, I'm in a jackpot," Abbruzzi said. "The guy thinks that me and Joey sold it."

It was, ironically, the kind of scam Merlino routinely pulled. But in this case there was no money, just a trailer-load of baby formula sitting in the parking lot of a Jersey Turnpike rest stop.

When Merlino said they were going to get another trailer shortly, Previte cautioned, "Do me a favor, please. Look in there first. . . . No more of that fuckin' baby food."

The swag deals would continue for nearly a year. Eventually Merlino would hook Previte up with a contact in Boston, and the negotiations would get even more bizarre. Previte would just shake his head and roll his eyes. He wondered sometimes how these guys ever made any money.

Previte's disdain for the organization was professional. Again and again he saw Merlino and those around him squander opportunities to make money because of their inability to plan, to think, to anticipate. All they knew how to do was grab.

For weeks Merlino talked about a load of nickels—"a half-million dollars' worth of nickels," he said—that was supposed to be stored in one of the trailers at a cargo yard along Delaware Avenue. Merlino figured they could unload them for half price, get maybe two hundred to two hundred and twenty-five thousand dollars. He said there were guys in the check-cashing business who would be in the market.

Previte was skeptical but played along. Sure, he said, he'd take them.

Unloading, however, could be a problem.

"You got five hundred thousand dollars' worth of nickels," he told Merlino, "you got a hundred thousand pounds."

It didn't matter. The hijackers never found the tractor-trailer with the nickels. It was just more talk.

There would, however, be trailer loads of toy trains, televisions, and computers. Most of those Merlino and Abbruzzi were able to move without Previte's help.

Like the meth transactions with Natale, Santilli, Rubeo, and Constantine, the swag deals Previte conducted with Merlino and Abbruzzi were slam dunks, solid criminal charges backed by tapes and evidence.

The feds, however, wanted more.

Previte had a load of stolen jewelry—valuables confiscated in other FBI operations—that he brought to Merlino. Skinny Joey said he had a jeweler friend in the market for hot stuff. Before he brought the stash to his friend, however, Merlino went through the pile, picking out women's jewelry, especially earrings. He took a few diamond-studded sets before bringing the rest to his jeweler. The jeweler plucked the diamonds out of most of the pieces, melted down the gold, and offered thirty thousand dollars for the lot.

"A nice score," said Previte, who hoped to split the take with Merlino.

It never happened.

Previte also came up with another Rolex—this one for Borgesi, who paid a thousand dollars for it. The watch, he said, was a gift for his brother.

"I'm doing all kinds of things with these guys now," Previte recalled. "Ralph's away. Joey's in charge, and I'm just trying to get close. . . . It wasn't that hard. Just like with Ralph, I had to bring 'em money."

In August 1998 Borgesi came to Previte with a proposition. He wanted to know if Previte would be interested in taking over his book. Joey had already given him the okay, Borgesi said. They wanted to have things lined up before the start of the football season.

"In the summer you generate about one-tenth of what you do during football," Previte later explained. "People bet baseball, but it was what I called maintenance. You were just using the income to pay the bills. Football is the mainstay of any gambling business. It's just so unbelievable how much money football generates."

Previte was pretty sure Borgesi wasn't interested in a partnership, despite what he claimed. But he had to listen and play along. He was being set up, just like Casolaro.

Borgesi said he had about forty regular gamblers whose action he was ready to turn over to Previte. He said that Bobby Luisi, a Boston-based wiseguy who had begun hanging with the Philadelphia mob, would also call down with action every week. Borgesi said he wanted his brother, Anthony, and his associate, Fat Angelo Lutz, to work with Previte, to get a share each week. Borgesi said he would be an inactive partner, taking a piece but leaving the bulk of the business and the work to Previte.

"I'm not gonna lie to ya," said Borgesi. "We can win thirty, forty [thousand] one week. We can also lose thirty or forty."

In that conversation, recorded on August 6, 1998, he told Previte that he was tired of all the aggravation that came with the business and that he'd rather just give it up. "I like to bet," Borgesi said. "I ain't a bookmaker, I'm a fuckin' degenerate. You know what I mean?"

Previte knew exactly what he meant.

"I didn't trust any of them," he said later. "I figured they were setting me up. But I didn't care. The FBI was backing my book. It wasn't my money. I couldn't lose. If they were gonna guzzle anybody, it would be the FBI."

But he had to let Borgesi know that he wasn't a complete sucker, that he knew what might be coming. He made a passing reference to the $212,000 Merlino had bet into him and never paid.

"It ain't my business, I don't want to hear it," Borgesi said. "You know what I'm saying. I mean, nothin' against Joey . . ."

Borgesi had been around Merlino all his life. He knew the way he operated. He had seen him pull scams and grab money. It was just the way he was. He was never going to change.

"You should see him when he's fuckin' bombed," Borgesi told Previte, laughing. "He's fuckin' the devil. Like there's fuckin' horns growin' out the back of his head."

Previte tried one more time to make the point, to let Borgesi know that he had his doubts. It didn't make sense that Borgesi would come to him out of the blue with this great business opportunity, he said. "Just between me and you, why me? What did I do to deserve it?"

He never got a straight answer, but it didn't matter. He was going into the loan-sharking business with John Ciancaglini, a top Merlino associate, and now he was going into the bookmaking business with Borgesi, a close friend of Merlino's and the crime family's new consigliere.

"I never liked Georgie," Previte now says. "In fact, of all those guys, I probably liked him the least. I thought he was a bully and a sneak. He liked to talk tough, but he was a punk."

Angelo Lutz, on the other hand, was a sucker who was regularly used and abused by guys like Borgesi. Previte felt sorry for him. And a part of him genuinely liked the roly-poly bookmaker.

Fat Angelo had seen several legitimate businesses fail because of his gambling habit. Anyone who knew him said he couldn't be trusted around money. But he was a likable guy, funny, engaging, a quintessential South Philadelphia bon vivant and raconteur. To sit down to a meal with Fat Angelo was to spend three or four hours eating well and laughing continuously. Not since the legendary Nicholas "Nicky Crow" Caramandi had the mob had such an accomplished con man. Fat Angelo, with a wink and a smile, could talk his way into or out of almost anything.

Borgesi referred to him as "Liar, Liar," a nickname that dated back to their childhood days on the corner. Even then, you never knew when Fat Angelo was telling the truth. But most of the time you enjoyed the story.

Lutz was an excellent cook; at one time he'd even run a successful catering business. He was also a surprisingly good dancer and an accomplished musician. Among other things, he marched every year in the famous Mummers Parade with one of the South Philadelphia string bands.

The parade, which runs up Broad Street on New Years' Day, is a five-hour extravaganza of sights, sounds, and buffoonery, each string band or fancy brigade trying to outdo the other with its elaborate costumes, musical themes, and over-the-top skits. In 1997 Fat Angelo was the hands-down crowd-pleaser, strutting his stuff as the Golden Buddha, the centerpiece of a musically themed skit called "Singapore Surprise" performed by the Italian-American String Band. He marched that day in nearly subfreezing weather dressed in flowing bronze-colored pantaloons, his bare head, arms, and rotund chest and belly painted gold and jiggling to the beat.

The crowd went wild.

Through most of the fall of 1998, Previte met with Lutz each week to settle up, going over the numbers from the previous weekend to determine who owed what to whom. From the way Lutz talked, Previte knew that he understood the bookmaking business completely. Previte also recognized that Lutz and Borgesi were scamming him.

By the end of September, about five weeks into the football season, one of Borgesi's bettors was already forty-six thousand dollars in the hole. Several others were down by ten to fifteen thousand.

Fat Angelo insisted they were trying to collect. In fact, he said that both he and Borgesi had spoken to "Beef," the gambler who owed the forty-six grand, and he had promised to come up with at least ten grand shortly.

Previte knew what was going on, but he let Fat Angelo talk. "Liar, Liar" was on a roll. "The kid's panicking," Lutz said of the mysterious Beef. "Seven years ago he went, he took a ride with Georgie to do something, to pick up money from somebody, and the guy was jerking Georgie off. . . . He owed him like ten thousand."

They met at a small restaurant, Lutz said.

"So, this is funny, we're sittin' there, Georgie's drinkin' an iced tea. So he, the kid, hands him fifteen hundred. Georgie turns around and says, 'What's this?' The kid goes, 'Oh, it's all I have.' This is after the kid had told him to come meet him and he would give him the ten.

"As soon as he gets the money, Georgie goes, 'Here, you keep it.'" Borgesi calmly handed the kid back his fifteen-hundred dollars, Lutz said. "Then he got the iced tea bottle and he cracked it over the fuckin' kid's head. Fuckin' blood was gushin' all over."

Had that really happened, or was Fat Angelo just playing him? Previte didn't know, nor did he especially care. Lutz's point was clear: Beef wouldn't be a problem. Fat Angelo assured Previte that he was good for the money.

"Our friend [Borgesi] says he's responsible for it," Fat Angelo said.

That, Previte knew, was pure con.

Later, after Previte's cooperation with the FBI was disclosed, Borgesi joked about the attempted scam. He admitted frankly that he and Lutz were betting into Previte's book, that Beef and half a dozen other bettors who called in each week—some giving their names, others merely their initials—were in fact placing bets for Borgesi and Lutz.

"We were trying to guzzle him," Borgesi said, claiming that he always suspected Previte was a cooperator. The only problem was, he said, "We could never pick any winners. We lost every fuckin' week."

Angelo Lutz also became an unwitting source of information for Previte. An inveterate gossip, he showered Previte with details about life within the Merlino faction. Early in October, for example, he confided that the consensus within the organization was that Ralph Natale was never coming home. Natale had been sentenced to sixteen months for violating parole, but Lutz said he was hearing that that was just the beginning of his problems.

Previte already knew that Merlino and the others had written Natale off. Among other things, they'd stopped sending money to Natale in prison, stopped sending cash to his wife, and stopped supporting his girlfriend. ("It was like I had died," Natale said later.)

"The thing I heard, between you and me, is that they got him [Natale] in a restaurant talking about shit," Lutz told Previte, during a discussion about the problems they were having collecting. "And within these next sixteen months, something else is coming up."

Previte was happy to reinforce that impression. "You can take that to the bank," he said.

Fat Angelo complained about Natale's management style, claiming that he was never able to get anything going during the four years Natale was boss. Natale "was a salesman," Lutz said disdainfully. He told Previte how he and Borgesi would go to meetings where Natale would boast about how he was going to put him back in the catering business, how he was going to get in the food supply business, how he was

lining up meetings with food and produce distributors. Lutz
said he would come out of a meeting with Natale pumped up
and ready to go.

"I swear to God, when you got done, you were ready to
. . . go out and attack the world. And then, nothing would
come of it."

Lutz said he and Borgesi opened Pasta, Cheese & Things
on their own after they got tired of waiting for Ralph to come
through with food-industry connections he claimed to have.
Previte said he understood completely. "If I tell you the
things he bullshitted me on," he said to Lutz. "You remember
The Exorcist? Your head would spin."

Lutz also complained about Natale's girlfriend, claiming
that the boss had called him several times at two or three in
the morning, asking Lutz to come pick him up at Ruthann's
place in South Philly and drive him back to Jersey.

"Two in the morning he'd ring the phone here. 'Ange, it's
me.' I'd have to get out of bed, pick him up, take him back."

Lutz also described an incident in a bar when Natale nearly
had someone stabbed because of a dispute over Ruthann. Previte countered with another rendition of the Viesti story.

"You don't have to worry about him no more," Previte
said of Natale. "New era. New beginning."

Lutz assured Previte that his past ties to Stanfa hadn't
hurt him in Merlino's eyes. "That guy loves you, believe
me," he said.

Previte knew, of course, that that "love" was based on his
ability to generate cash. And he also knew that as long as the
FBI was footing the bill, he wouldn't have a problem doing
just that.

They gossiped about other members of the organization, including a few associates who had disappeared during the trouble with Stanfa but now were back in town and
happy to do business with Merlino again. Lutz, who had
been on Stanfa's hit list and who hadn't run away, didn't

understand why Joey would have anything to do with them.

Previte, on the other hand, knew that loyalty for Merlino was a cash-and-carry proposition. You brought the cash, and he'd be happy to carry you.

14

One mobster who was happy—indeed eager—to bring Merlino cash was Bobby Luisi.

"Boston Bob," as Borgesi called him, was introduced to Natale and Merlino a few months before Natale went away on the parole violation. A friend from New England who had been in jail with Natale made the introduction. He told Natale that Luisi had gotten caught in the switches in Boston, and was unhappy with the way things were playing out up there. He was interested in doing some business with the guys from Philadelphia.

The Boston branch of La Cosa Nostra might be the only crime family in America that is more dysfunctional than the Philadelphia mob. It has been torn apart by years of internecine bickering—bloody squabbles that have left dozens dead—and by an undercover operation that brought shame and embarrassment not only to underworld figures but also to the FBI.

The Federal Bureau of Investigation has had a staggering record of successes in its fight against La Cosa Nostra over the past thirty years. Once the bureau decided that the mob did, in fact, exist (for years J. Edgar Hoover had denied that the Mafia was operating in America, preferring to have his G-men chase bank robbers), it set about the systematic dismantling of the organization. Robert Kennedy, the RICO statute, Joe Pistone (aka Donnie Brasco), the Commission Trial,

Rudy Giuliani, and the Pizza Connection are all part of the Bureau's legendary decade-long assault on the mob.

But with each success came an increase in arrogance. Ask any big-city police detective or state police investigator who works the organized crime beat what he or she thinks of the FBI, and the answer will be laced with expletives. It's always a one-way street with the "special agents": In their view, they have all the answers. They control all the resources. They are the only agency capable of making the serious and meaningful cases.

There is a culture of distrust, a general disinclination by the FBI to share information with other investigative agencies. It is organized crime and disorganized law enforcement. Taken to its most deadly and damaging extreme, it is that kind of smug arrogance and righteous posturing that opened the door for the American tragedy now simply referred to as 9/11. The bombings of the World Trade Center and the Pentagon by Arab terrorists may not have been avoidable. But what was clear in the aftermath is that the FBI in New York and Washington ignored or refused to take seriously warnings and questions being raised by its own agents in lesser field offices. These field agents weren't "in the loop," according to the total control mind-set that pervades the Bureau, and thus it was assumed that they couldn't be in a position to know anything of significance.

That same attitude has been a part of the fight against organized crime for two decades. It would manifest itself in Philadelphia in the decision to cut a deal with Ralph Natale—make a deal with the devil—in an ends-justify-the-means move that had less to do with justice than it did with making high-profile cases.

Want to turn an FBI agent's face red? Mention any one of these three names: Salvatore "Sammy the Bull" Gravano, Gregory Scarpa, or James "Whitey" Bulger.

Gravano's story is perhaps the best known. A ruthless mob underboss who admitted to nineteen murders, he cut a deal and got out of jail in five years after agreeing to testify against his former friend, mob boss John Gotti.

Sammy got a book and movie deal, took millions with him, and moved to Phoenix, where he thumbed his nose at the mob and, eventually, at those who had cut him one of the biggest breaks in underworld history. Not content to live a life of quiet wealth and prosperity, Gravano, along with his wife, son, daughter, and son-in-law, got involved in the drug trade out west. He became a major supplier of Ecstasy, the party drug that began spreading in epidemic proportions among teenagers and young adults in the 1990s. In 2002 he was sentenced to twenty years in prison after pleading guilty to federal drug charges. He got a similar sentence in a related Arizona State narcotics case.

The question that had always plagued the families of Gravano's nineteen mob murder victims was how much was convicting John Gotti worth? Now that same question can be asked by the families of anyone whose life was ruined by the drugs Gravano peddled after the FBI gave him a new life.

Gregory Scarpa is dead. The Brooklyn-based wiseguy died of AIDS, contracted from a tainted blood transfusion while he was being treated for bleeding ulcers. But for several years, during a bloody Colombo family mob war in New York in the early 1990s, Scarpa was a "top-echelon" FBI informant—the highest rank the bureau places on a cooperator.

Scarpa was feeding information about the mob war to his FBI contact. But other mobsters and their defense lawyers have since charged that the FBI was also feeding information to Scarpa. In fact, they allege that Scarpa used that information to set up and kill several rivals during the mob war.

It was not the FBI's finest hour. Several convictions have been overturned because of the Bureau's failure to disclose Scarpa's role as an informant in cases where the feds used his information to get indictments or to convince others to testify.

Finally, there is Whitey Bulger. For more than twenty years, Bulger and his top associate, Stephen "the Rifleman" Flemmi, had carte blanche in the Boston underworld because they were supplying their FBI contacts with information about the Patriarca crime family.

In essence, the feds made a decision: deal with the Irish gangster to get the Italians. The FBI's point man in the operation, John Connolly, who had grown up in the same Southie neighborhood as Bulger, was recently convicted of racketeering charges for his role in the Bulger affair.

Whitey (whose brother William has served as president of both the Massachusetts State Senate and the University of Massachusetts) went on the lam shortly before he and Flemmi were indicted in 1995. He fled, many believe, because he was tipped in advance that the indictment was coming. He has graced the FBI's 10 Most Wanted List ever since. Cynics now wonder how hard the Bureau is trying to find him. A Whitey Bulger in custody and willing to talk about all he did while working for the FBI would be a lot worse for the Bureau than a Whitey Bulger on the run.

The topic of several federal and judicial investigations, and the focus of the riveting book *Black Mass* by two *Boston Globe* reporters, the Bulger fiasco was the FBI at its arrogant and self-assured worst.

Bulger and Flemmi used the feds to get rivals out of the way so that they could expand their own rackets, which ranged from drug dealing and extortion to gunrunning and murder. As in the Scarpa case, many defense lawyers and jailed mobsters now claim that their convictions were improper and should be overturned. And as in the Scarpa case, sources claim that Bulger and Flemmi used information provided by the FBI to kill underworld figures, including some who were attempting to implicate them in crimes.

Most of this is detailed in a 600-page report issued in

1999 by U.S. District Court Judge Mark L. Wolf. The report was based on eighty days of hearings and testimony from forty-six witnesses, including FBI agents and mobsters.

Flemmi, who was convicted in the case that sent Bulger scurrying to parts unknown, claimed that they were given an underworld pass by the FBI. He said he and Bulger were told, "You can do anything you want as long as you don't clip anyone."

Judge Wolf noted later in his report that the FBI continued to deal with Bulger and Flemmi even *after* they became suspects in a series of gangland hits. The judge found that the two wiseguys were clearly using the Bureau to advance their own criminal enterprises.

One anecdote included in the judicial findings involved Bulger's reaction in 1985 after learning that the DEA had targeted him and had bugged his automobile. Bulger ripped the bugs out. DEA agents, hearing what was taking place, rushed to recover the tiny high-tech electronic surveillance devices, which were valued at between fifteen and twenty thousand dollars apiece.

When they confronted Bulger, he told them they shouldn't be bugging him. "We're all good guys here," he said. "You're the good good guys and we're the bad good guys."

It's a funny story. Except to the family members and friends of the people Bulger killed.

Murder, of course, was a way of life in the Boston underworld, just as it was in Philadelphia. Bobby Luisi had lost a father, a brother, and a cousin in a notorious gangland hit in a restaurant just outside of Boston in November 1995. The shootings came in the midst of a turbulent period of unrest in which two factions, one headed by Francis P. "Cadillac Frank" Salemme, vied for control of the Patriarca crime family.

In certain underworld circles, the fact that Robert Luisi Jr.

did nothing to avenge those murders said all anyone needed
to know about the erstwhile wiseguy.

Not for nothing, said several Boston area underworld
sources, did every faction involved in the mob up there balk
at the idea of formally initiating Luisi into the crime family.
He was a hustler, a shyster, and a sometime drug dealer who
talked a better game than he played.

Perfect, in other words, for an organization headed by Natale and Merlino.

Bobby Luisi stopped backing Salemme when the mob
boss refused to make him a member of the crime family.
Among other things, the FBI has linked Luisi to at least one
murder and eight other murder conspiracies, most of which
stemmed from the power struggle for control of the Patriarca
organization.

After he split with Salemme, Luisi set up his own operation; federal authorities would later charge that he was involved in gambling, extortion, and cocaine trafficking,
primarily in Boston's North End, but also in certain outlying
suburban areas.

By 1998 Luisi was looking for another alliance, someone
who would back his move in Boston, who would give him
the status and prestige he couldn't get from the seasoned
wiseguys in New England who had watched him grow up
and who knew him too well.

"I met him sometime in the late summer or fall of 1998,"
Previte said. "I think Joey or Georgie introduced him to me.
They said he was from Boston and that he was going to be
with us."

Previte said he didn't know much about Luisi's background but was less than impressed by the stocky, dark-
haired forty-year-old.

"I'm a pretty good judge of people," he said. "Bobby was
a wannabe doofus. You could see that. He wanted to belong.
I knew I would be able to work him. When you have a guy
who will do anything to impress, it's easy to lead him down

the primrose path. I didn't tell the FBI that, of course. I told
them it might be difficult. But I knew it wouldn't be. You
could just see it. It was going to be easy."

Based almost entirely on the recommendation of his
prison friend, Natale had welcomed Luisi into the fold. He
and Merlino discussed bringing the new guy along slowly.
Since they didn't know too much about the Boston area,
Natale said, they shouldn't rush into anything. He put
Luisi with Borgesi. At that point, Boston Bob was still a
mob associate.

By the time Previte met Luisi, though, Natale was already
in jail on the probation violation, and all talk of going slowly
with the Boston wannabe had long been forgotten. Luisi was
now a full-fledged soldier, and soon to be a capo in charge
of the Philadelphia crime family's first foray into New England. Natale's caution had been subsumed by Merlino's desire for cash.

Luisi had become a fast friend of Borgesi's. They and
their families—each had young children—took a vacation
trip to Florida together, visiting Walt Disney World. Later
they would also travel to Las Vegas, although on that trip the
families stayed behind.

Luisi had his own crew in Boston, four or five guys
who were involved in various scams. They were not, however, high-caliber operators. The tractor-trailer load of
vodka that they grabbed consisted of a brand called
Moscow that no one had ever heard of and that no one
was interested in buying. And the truckload of frozen
shrimp that Luisi kept promising Merlino in the summer
of 1998 never materialized.

Still, Bobby Luisi was coming up with cash each month
for Skinny Joey—ten grand. Luisi had bought his way into
the Philadelphia mob on an installment plan: he had promised Merlino a hundred thousand dollars in return for his
membership, to be paid in monthly installments of ten thousand dollars. The need to get that money—to show that he

was an earner, that he could produce—eventually led Luisi into a major cocaine deal. Previte served as the middleman; the buyer was an undercover FBI agent.

Those deals, beginning in the winter and ending in the spring of 1999, would be the last recordings Previte made for the FBI. But before he reached that point he would tape dozens of other meetings with wiseguys in his own backyard.

Money, murder, and broads dominated most of those conversations. But in one, John Ciancaglini provided a poignant assessment of the price he and his family had paid for their involvement in the mob.

He and Previte met in the Oregon Diner on November 2, 1998. It was the meeting Merlino had been trying to set up since the summer, the meeting where the two were to shake hands and put the past behind them.

"He was a smart kid," Previte said of Johnny Chang, who at forty-four was nearly as close in age to Previte as he was to Merlino. "He wears a sports jacket and he's got those Mister Peabody glasses, but he was a real gangster. He knew what it was about. And he knew what Joey was about."

While Johnny Ciancaglini has consistently denied it, the feds contend that he was forced to make a heart-wrenching underworld decision when he was released from prison in January 1995. He could either side with the beleaguered and imprisoned John Stanfa against the Merlino faction that had gunned down his brother Joey. Or he could side with Merlino against the Stanfa mob that had killed his brother Michael.

The feds, of course, say that Johnny Chang chose to go with Merlino.

There are others, however, who believe that Ciancaglini got caught in the middle of something over which he had no control, that through it all he was just trying to survive. While authorities charge that he got involved in the book-making and extortion rackets, the record also indicates that he opened several legitimate businesses. He was a contractor, building homes in South Philadelphia. He had an exterminating business, and another in which he silkscreened T-shirts for companies, stores, athletic teams, and the like.

Unlike Merlino and many of those around him, Johnny Chang wasn't afraid to work. More to the point, he knew how to make money. He and Previte were alike in that respect. Over coffee at the diner that afternoon, he explained to Previte why he was still angry at Previte for extorting one of his bookmakers during the Stanfa era. He also said he was upset that Natale hadn't backed him when he first complained about the extortion and asked that the ten thousand dollars be returned. That was the source of the friction and animosity that he had toward Previte. It wasn't about the money, Johnny Chang said, it was about the principle.

Previte said he understood. But he also pointed out that he never saw any of the money. He was collecting for Stanfa, who as usual put the cash in his own pocket. "I didn't get nothing out of it," Previte said. "Can you put yourself in my shoes?"

Both agreed that it was nothing personal, that they'd been on opposite sides in what proved to be a deadly game of mob one-upmanship. The extortions and shakedowns led to shootings and murders. And in that regard, Johnny Ciancaglini had clearly paid the biggest price.

"I'm considered in the middle," he said, " 'cause I had one brother on your side and one brother on the other side. . . . I'm just saying that what happened wasn't right."

"I'm sure of that," said Previte.

Each apologized to the other and said they looked forward to working together.

They ended the conversation on a cordial note. Previte talked about his bookmaking operation. Ciancaglini mentioned the homes he was building, some of his other ventures, and his overall philosophy of making himself scarce.

Previte concluded by telling Ciancaglini he was sorry about what happened to Michael.

"My brother Michael was an ace," Johnny Chang said of his youngest brother. "The only thing about Michael, he had a head like a rock and he didn't take no shit."

"Tough kid," said Previte.

"He didn't take no shit. I told him all the time, wait till I come home, Mike. I got seventeen months. Wait till I come . . . home."

Egged on by Natale, Michael Ciancaglini couldn't wait. It cost him his life.

Not all of Previte's conversations were as somber or philosophical. A few weeks later, he and an associate got into a debate about women. The associate, identified on the FBI transcript only as "unidentified male," had a date that night. (Since the topic was irrelevant to any investigation, the feds spared him the embarrassment of identification.)

"Your thing still working?" Previte asked.

"Yeah, you kiddin'? Like clockwork."

They then discussed a mutual friend who had given up "the pump" because it never worked right. Instead, he found a more effective method of enhancing his performance—Viagra. The pills, the associate said, went for "eleven dollars apiece."

"On the street?" Previte asked.

"No, the pharmacy."

"That's a good scam," Previte said. "That's what we oughta do. Get the packages. Make 'em up."

The associate said he had read about some guys "from Mexico" who were smuggling the stuff into the country. "They had two hundred and fifty thousand Viagras coming over the border," he said.

"Were they real?"

"Yeah. Comin' over the border. That's a good thing to get into now. You oughta do it."

There were all kinds of options, he insisted. In addition to tablets, he said he had heard "now they're making one outta chewing gum."

Previte was incredulous. "Outta what?"

"Outta chewing gum. For women. Women and men."

"Chewing gum. Get the fuck out of here."

"No, it'd be okay. You could be hittin' it and blowing bubbles."

"Ah, they'll come up with anything. They'll come to you with anything."

"You know you can go for two days with them things. You think you're twelve years old, man."

The only possible problem, he said, was "you go blind."

"Blurry eyes never killed nobody," Previte said.

Then the associate admitted he had tried Viagra himself. "I was in Margate. I popped one in Margate. By the time I got to Hammonton [a thirty-minute car ride] I had to make it raw jerking off."

Drugs of a different sort would be the focus of Previte's undercover work for the next six months. Cocaine would replace Viagra as the topic of discussion on many of the tapes he made.

He planted the seeds for the deals in a conversation he had with Luisi in November, during one of Boston Bob's many trips down to Philadelphia. Previte said he had a "friend" in the Boston area, a guy who was a "real good earner." He said he had done business with him when he was in Philadelphia, and told Luisi "we made a lot of money together." The guy had moved up to Boston two or three years ago, after he and his wife had divorced; the wife had taken the kids, and he transferred his operation up there to be close to them.

It was all part of the cover story.

The guy was in the import-export business, Previte said, with a company named, "Irish something or other." Did a lot of stuff out at the airport, and was always interested in side deals. Maybe, Previte suggested, he could hook him up with Luisi.

Anxious to impress, Luisi said he'd like to make that connection. Told it might be a chance to make money, Merlino told him to go ahead. Whether Joey knowingly approved the drug deals that followed would become a point of contention still debated in law enforcement and underworld circles alike.

Merlino, through his lawyers, would later insist that he had no idea Luisi and Previte were cooking up cocaine deals. He thought they were talking about stolen property, he said. Most of the tapes in which he discussed Boston deals with Previte are vague enough to justify that spin.

On January 6, 1999, before his first trip to Boston, Previte had a lengthy discussion with Merlino about stolen property deals, the problems they were having getting trucks to move the swag, and the fact that they both wanted to make more money.

Previte said his contact in Boston was a guaranteed source of cash, that he could come up with some kind of deal every week. Previte insists the conversation swung from stolen property to drugs, and that Merlino understood. But the words on the tape are less clear.

"This, this is big up there," Previte said at one point, putting his finger to his nose. It was clear to them both that he was talking about cocaine, he says, but the tape and transcript only include the words, not the hand gesture.

"Just make sure that what's-his-name [Luisi] can meet me up there to meet the guy," Previte said.

"That's what I'm sayin', yeah," Merlino said.

Both Previte and Luisi would claim they moved forward with the drug deals after getting the okay from Skinny Joey.

The man who made it all happen was FBI agent Michael McGowan, a veteran undercover operator who played Luisi and his Boston crew like a violin. Before the cocaine sting was completed, McGowan, working with Previte, had both

audio- and videotape of the drug deals being set up, the cocaine being delivered, and the cash being exchanged.

It was even better than the meth case against Natale.

McGowan posed as "Mike Sullivan," the operator of a
company in downtown Boston called "Irish International"
that was involved in the import and export trade. He did a lot
of work, he said, out of Logan Airport. His office, in a modern professional building near the Marriott Hotel, was
bugged for sound and wired for video. Every meeting that
occurred there ended up on tape—Luisi setting up the deals,
Luisi and his associates making the deliveries and collecting
the cash.

McGowan brought just the right combination of attitude,
edge, and respect to his meetings with the erstwhile
wiseguys. "This guy knew what he was doin'," Previte said
of McGowan. "He and I worked perfect together. It was like
we didn't have to talk about anything, we just played off one
another. He was smooth. He understood."

Previte was just as effective, playing to Luisi's vanity despite what he thought of him personally. The combination
had Boston Bob's head spinning.

Previte made his first trip to Boston on January 11, 1999.
There he met with Luisi, who introduced him to the members of his crew. He then took Luisi to the offices of Irish International for a get-acquainted session with Mike Sullivan.
On their way to the meeting, Previte again told Luisi that
Sullivan was an earner. He also let Luisi know that Sullivan
sometimes dealt in drugs.

"He might be interested in this stuff," Previte said, pointing his finger to his nose.

Luisi nodded.

The next day Luisi and Sullivan met again and agreed to
do business. Previte, acting as the middle man, told Sullivan
that Luisi was "in charge of everything here in Boston." Sullivan could deal with Luisi, Previte said, the way he had
dealt with Previte in Philadelphia.

"You're with us," Luisi told Sullivan, who smiled and shook hands with the wiseguy.

Then they all headed out to lunch in the North End. They went to one of Luisi's favorite spots, the Caffe Vittoria. He loved the atmosphere of the Hanover Street coffee house, a combination of old-world charm and hip bistro chic.

Bobby Luisi considered himself a player.

And he wanted everyone to know it.

15

Like Natale in Philadelphia, Luisi had big plans for the Philadelphia mob in New England. It was his ticket to the top. With Cadillac Frank Salemme and most of his chief associates in jail, Luisi saw a power vacuum. Who better to fill it than himself?

But like Natale, he had trouble turning his plans into reality.

His first three deals with "Mike Sullivan" were busts.

Luisi took five fur coats that Sullivan said had been stolen in Connecticut. They were worth about thirty thousand dollars. No problem, said Boston Bob. But he ended up selling one for six hundred dollars, giving another to his wife, and a third to a relative. He then sheepishly returned the other two.

A month later, he and an associate picked up ten cases of stolen film—Kodak 110 and Polaroid 600—that Sullivan said "fell off a truck." Again, no problem, said Luisi. Two weeks later, he returned the film. Couldn't find a buyer.

Finally, for nearly two months, in the midst of the deals for the coats and the film, Sullivan pitched a diamonds-for-cocaine transaction. He said he had a contact who wanted to unload a stash of hot diamonds, and he'd be willing to trade them for coke. His contact was looking for "three bricks," three kilograms of cocaine. Luisi said a kilogram was going for about thirty-seven thousand dollars. Sullivan didn't bat an eye. This time, he said no problem: his guy had enough diamonds to cover that cost. Luisi promised to set the deal up.

It never happened. Still, Sullivan kept trying. To further whet Luisi's appetite, the FBI undercover handed him a Rolex watch—part of the stash of hot jewelry, he said. He wanted Luisi to have it. It was an Oyster Perpetual Date Submariner with a blue bezel and dial, 18-karat yellow gold and stainless steel. It was worth about five grand. Luisi loved it. Sullivan said his contact had a dozen of the watches and was looking for someone to move them.

In a six-month period, the FBI had provided Previte and Sullivan with three of these Rolex watches for Merlino, Borgesi (for his brother), and Luisi. Combined, they were worth between fifteen and thirty thousand dollars—a small price to pay for the access it gave the feds to an organized crime family that, in many other respects, didn't know what time it was.

Things were also sputtering for the Merlino organization in Philadelphia. The stolen property ring that Skinny Joey and Ralph Abbruzzi had been using to generate cash had cut back on its operations; internal bickering and forced profit sharing with the mob had taken a toll. Eventually several lower level members of the ring would turn informant. One, in fact, was discovered trying to tape a conversation with Frank Gambino, another Merlino soldier involved in the swag trade.

Gambino, a sixty-seven-year-old wiseguy who had spent nearly half of his adult life behind bars, discovered the tape during a card game. The informant, a guy named Freddy Angelucci, got into a scuffle with another card player. As they wrestled, a tape recorder Angelucci had in his pocket fell on the floor.

Almost in unison, Gambino and several other players shouted, "You're wired!" Angelucci, deciding an explanation would be pointless, bolted out of the apartment. Gambino proceeded to stomp the tape recorder to pieces.

But the damage had already been done. Angelucci had recorded dozens of conversations, including one in which Gambino offered these words of wisdom: "The least people that know your business, the better off you are."

By that point, in early February 1999, the feds knew more about Gambino's business than the aging wiseguy could imagine. Angelucci, who said he had been threatened, robbed, and cheated, eventually testified for the government. From the witness stand he offered a less-than-flattering description of the way the Merlino mob horned in on the stolen property ring, first demanding a street tax, then insisting on a piece of each stolen tractor-trailer load.

"They were rip-off artists," he said of the mobsters. "They always came up funny with the money."

Previte knew exactly what Angelucci meant. Despite all the talk, he hadn't seen a lot of money from the swag deals. As a result, he had to act upset. He had had to share the profits from the bicycle deal with Merlino and came away from the scam with less than five grand. He'd also been burned on the "stolen" jewelry he had given to Merlino. Merlino had told him his jeweler, who melted all the stuff down, thought it was worth between twenty-five and thirty thousand dollars. But the only cash Previte saw from the deal was two grand Joey handed him a few weeks after he delivered the stash of jewels.

"I figured they were robbing me again," Previte said. "But at that point the idea was just to keep the investigation going. So I would bitch and moan, but I didn't really care. This was all the FBI's stuff anyway. I heard later that Joey didn't get too much of the money either. Joe Ligambi knew the jeweler and ended up with most of the cash. That's what I heard, anyway. They all steal from one another. That's the way they are."

At the time, both Previte and the FBI were more interested in the cocaine deals. And that's where he shifted his attention. Luisi had a history of dealing coke, but was being very cautious about putting a deal together for Irish International.

Realizing Merlino was strapped for cash, Previte decided he could jump-start the cocaine transactions by getting Merlino more directly involved. It was clearly what the FBI wanted when the sting was set up. Luisi might think he was a player, but he would hardly be considered a major catch for the feds. Merlino, on the other hand, had become the primary target of the Philadelphia office. He was even more important than Natale.

On March 25, before his latest trip to Boston, Previte met with Merlino and quietly complained about Luisi. "I gotta go back up there," Previte said. "'Cause he didn't want to do the other thing, but now he wants to hook somebody else up with my guy, whatever."

"Well, tell him to fuckin' do what he gotta do," Merlino said. "I'll tell him."

After a few minutes discussing local business, Previte shifted the conversation back to Boston and the diamonds-for-cocaine deal that Luisi had blown. He also held out the prospect of more Rolex watches, an item he knew Merlino was interested in.

Luisi, tan and relaxed, came back from a trip to Walt Disney World with Borgesi and his family at the end of March, but it would take another month and more prodding from Previte before any cocaine ended up on the table.

Boston Bob was picking up mixed signals. He had traveled down to Philadelphia at least twice earlier in the year to discuss business with Merlino. During one of those trips, he attended a Valentine's Day party where he had a brief discussion with Previte and Merlino about the deals that could be set up in Boston. Previte insists that Merlino gave the okay that night to Luisi, that Boston Bob had Merlino's approval to set up the cocaine deals with Previte's guy. But Previte wasn't wearing a wire that night—since he was there with a date, the FBI decided it would be too risky—and defense attorneys would later use the fact to undercut his contention that Merlino approved the deal.

Merlino may have given the okay, but Skinny Joey was always cautious. He said things indirectly. His conversations were governed by a simple dictum: one word is better than two, a nod is best of all.

"Help that guy" might have been the extent of what Merlino said at the party. To Previte it was obvious what he meant.

But Luisi was still feeling his way around the Philly mob. And at this same time, he was getting a clear signal from Borgesi *not* to deal drugs.

Borgesi, the young consigliere, may have been many things—a bookmaker, an extortionist, a bully—but he wasn't a drug dealer, and he didn't want to be around anyone who trafficked in narcotics. Maybe it was the fact that he had two young kids of his own. Maybe it was because he had seen too many guys in the neighborhood destroyed by cocaine, heroin, crack, and meth. Maybe it was just that he knew a narcotics pinch carried a lot more weight. Or strangely enough, maybe he just couldn't stomach the thought of facing his family—his mother, his brother, his wife and children—if he were ever charged with doing something as despicable as dealing cocaine.

To many, Borgesi was ruthless, arrogant, and egotistical. But the thirty-five-year-old wiseguy was also plagued with a conscience. He brooded constantly, always concerned about who was being investigated and for what. He described himself, only half jokingly, as a "nerveen," a worrywart. There were times when the anxiety and tension, brought on at least in part by his personality, would result in crippling migraine headaches, forcing him to a darkened bedroom for hours until the pounding stopped and the nausea receded.

Investigators saw Borgesi's behavior as a sign of weakness. They would hear him ranting and raving on a wire-tapped phone, complaining about some investigation or news report, and would be convinced that he was a weak link, a logical target, someone who could be co-opted and convinced to cooperate. Borgesi had never done any time; all he had on his sheet was one minor gambling conviction, and there were those in law enforcement who insisted that the prospect of a heavy prison sentence would send the young mobster scurrying for the security of the Witness Protection Program.

It was another case of the feds misreading one of their targets. The only thing that would shame Borgesi more than dealing drugs, said those who truly knew him, would be the prospect of cooperating with the FBI.

"He'd kill himself first," said one former associate.

Borgesi, who had attended Central High School, arguably the best public secondary school in the city, loved to debate the issue of "the mob." What was it? Who was it? Did it even exist?

"Who am I?" he would say as he sat eating lunch in a South Philadelphia restaurant, joking about the latest organizational chart, asking where he was ranked in the crime family. "This is all bullshit. The cops don't know dick. They make stuff up about me and my friends."

He insisted he was never involved in any "stupid stuff," a euphemism he used for murder. And he railed against drug dealers and drug users.

"No way, on my kids' eyes, would I ever get involved with that shit."

Luisi, who had spent countless hours with Borgesi, heard the admonition many times. Now he was hearing something else.

By the end of April, Previte was pushing hard to get a cocaine deal in place. He told Merlino the money was just sitting there, but that Luisi was still dragging his feet.

"He's a good kid, but he ain't doin' what he said he was gonna do," he told Merlino in a conversation recorded on

April 27. "I got big money sittin' on the line. . . . Is there any way you could just tell him to do what he gotta do?"

Merlino promised Previte that he would. In fact, they agreed to have a three-way phone conversation once Previte got back up to Sullivan's office in Boston the next day.

"I'm tellin' ya, I got like forty, fifty Gs sittin' . . . if he does the one thing."

"Well, I'm gonna tell him," Merlino said.

"And I can bring you back ten."

"Right. I'm gonna tell him. I'll say, 'Bob, you know, help my friend.' I'll tell him. I'll say, 'Whatever he says, just do it.' That's all. What the fuck. Let's just make a bunch of money."

Money, as always, was a major concern for the young mob boss. In that same conversation, he told Previte that he was being audited by the Internal Revenue Service. Amazingly, Skinny Joey genuinely sounded surprised that anyone would question his finances. This from a guy who, during his rise to the top of the mob, had listed his "occupations" as coffee shop impresario, salesman for a home remodeling company, and sandwich maker in a steak shop, even as he and his family lived in a lavish $285,000 town house in an exclusive area of South Philadelphia and rented a condo at the Jersey shore each summer. He drove a Mercedes and also tooled around on a Harley-Davidson motorcycle. His wife favored Jeeps and Land Rovers.

The Mercedes, of course, was leased by the home remodeling company; for a time, Previte paid the lease on the Land Rover, with cash supplied by the FBI. The motorcycle belonged to Anthony Accardo, another mob associate. The town house was listed in Danny D'Ambrosia's name.

The feds, Merlino told Previte, had subpoenaed all kinds of financial records. They had visited two insurance companies from which he had purchased life insurance policies. They had pulled all the wage and salary records from a beer distributorship where Merlino's wife, Deborah, worked as a bookkeeper. They had even gone to Boyd's, a well-known Philadelphia clothing store where Merlino used to shop.

"I don't even go the fuck in there [anymore]," Merlino said.

Merlino's conversation with Previte offered the FBI some insight into just how much money he was generating in the underworld. Despite the fact that he was always broke and constantly looking for another score, Merlino had gotten his hands on a substantial amount of money. Even the amount he claimed on his tax returns—and no one believed that was an accurate reflection of the cash that had gone through his pockets—was more than enough to provide a comfortable living for most individuals.

"Me and my wife, in the last two years, I didn't pay them," Merlino said of his dealings with the IRS. "I only paid them like eight thousand. But I made, I made two hundred thousand. And my wife made . . . she paid, you know, paid hers. She made . . . three hundred thousand in two years."

Deborah Merlino clearly was a well-paid office worker. That she worked for a company controlled by a Merlino mob associate added to federal speculation that her wages were simply a way to launder and funnel money to her husband. She paid taxes on her earnings, however, and has never been charged with any income tax violations.

Merlino, in that cockeyed way he had of looking at any situation, saw the IRS probe as a positive. "It's a good sign, though, you know what I mean?" he told Previte.

As with the notification of the wiretaps at the Gianna's party, Merlino said, friends and lawyers had told him the IRS investigation might mean the feds had nothing else on him. "Like they said they [were] gonna get you for murder, extortion. Now they're goin' to this?"

Merlino dismissed the IRS as a minor annoyance, the cost of doing business. It almost sounded as if he would welcome a pinch, if all it involved were financial charges. In the meantime, he told Previte, "I want to start makin' some fuckin' money. Fuckin' this is ridiculous."

Previte agreed. And once again he said there was plenty of cash sitting in Boston.

"My guy's been waiting and waiting and waiting," Previte said, taking a not-so-subtle shot at Luisi. "Then he tells my guy, get him film. They get him all kind of film. They bring him the fucking film back. They tell him, get him the mink coat. They bring everything back. What the fuck . . . He brings them shit and they don't move shit."

But that was about to end.

The next day, from the offices of Irish International, with FBI agents by his side and the phone tap recording it all, Previte had a three-way conference call with Merlino and Luisi.

After a brief discussion about their various ailments—Merlino had pulled a muscle in his thigh playing softball the night before; Luisi and Previte were both fighting off colds and sinus infections—they got down to business.

As always, Joey was cryptic. "Bob," he said, "is that guy, ya know, do what he's got to do over there for him."

"Oh yeah," said Luisi.

"All right," replied Merlino.

"Yeah," Luisi said again. "That's, that's gonna, that's gonna be."

"Okay," said Previte.

"All right," said Merlino.

"Ya know?" added Luisi.

"You got it," said Merlino.

The conversation ended with Luisi agreeing to meet Previte in an hour at the Caffe Vittoria. If there was any doubt about what was discussed in the conference call, Luisi and Previte cleared it up during a "walk-talk" after they arrived at the café. Luisi showed up with two associates. Previte was there with Sullivan. Everyone shook hands. Then Luisi asked Previte to "take a walk."

Previte, wired and ready, was happy for the private time.

First Luisi apologized for the business with the fur coats. "Don't worry about it," Previte said.

Luisi then said he and his associates were waiting for their cocaine connection to come through. "I don't got ten cents on me, Ronnie," he said. "I'm waiting for this guy too. I got no money."

Previte and Luisi then talked about a long-term deal. Previte said his guy was "looking for two keys right now" and that that could become a regular order if things worked out.

"I'm sure once he gets moving he can keep doing that," Previte said. Both he and Sullivan would offer the same explanation. Sullivan was tapping into a suburban market— businessmen and professionals who worked in the city but who didn't want to buy off the streets and risk getting ripped off. Sullivan could exploit that market because he was one of them. They trusted him.

"He's got the cash for two keys right now," Previte told Luisi.

Luisi, in turn, said the cost had gone down. While he had originally told Sullivan that the cocaine might cost thirty thousand dollars a kilo, it was now being priced at "twenty-five apiece," he told Previte.

"You know, I don't know what they were really paying for it," Previte said later. "And I'll tell ya what, if this had been a real cocaine deal, there wouldn't have been another one. The product they delivered was weak. If they had stepped on it any more there would have been addicts all over the city getting attacks of diabetes."

Sugar was sometimes used as an ingredient to expand and dilute cocaine. The process, called "stepping on" the product, increases the quantity but reduces the quality. The profit margins grow as the product, pure at the point of origin, is expanded and diluted while it works its way through the market. Although he could only speculate, Previte suspected that Luisi and his associates were buying a kilo of cocaine from their supplier and then adding enough sugar to turn it into two kilos before delivering it to Sullivan. If they were paying twenty-five thousand dollars per kilo, they would turn a 100 percent profit on the deal.

Previte says he wasn't concerned about what Luisi and his crew were paying for the coke. Putting the deal together and getting it on tape, making the case against Luisi and Merlino—that's what it was about. The conversation during the walk-talk was one of the first in which there was little doubt about what was being discussed.

Luisi said he was reluctant to get involved but that he needed the cash. He had "obligations," an apparent reference to the ten grand he had to send to Merlino each month. "I'm having a hard time keeping up, you know?"

"I understand," Previte said. "I understand exactly what you're saying."

Previte then held out a carrot. He told Luisi that Sullivan could probably "do two a week for you . . . every fucking week. Big money for everybody."

Luisi nodded, then mentioned the party and the okay he had gotten from Merlino to move forward. "I don't want to mention no names," he said. "That other guy said help that guy out. . . . When he said that, that gave me the green light."

"I know."

"You know, that's what Joey wants."

"Well, that's what I say," said Previte. "We got the green light."

Both Previte and Luisi agreed that Borgesi would never have to know about what was going on in Boston. But Previte, ever the cynic, said that there was a difference between knowing and collecting. There was, he said, a certain hypocrisy in saying no to drugs but then accepting money that you knew came from drug dealers or a drug deal.

"You know, it makes me laugh," Previte said. ". . . Nothing against George. I love Georgie. But it's all right, nobody wants to know, but everybody wants to spend."

"Yeah."

"I say, 'Here's this,'" Previte said, making a gesture as if he were delivering an envelope full of cash. "Well, no. Okay, thank you. . . . And they make the fucking deposit slip."

Still, Luisi wanted to leave himself with some room to maneuver. He insisted that he wasn't really involved. All he was doing, he said, was putting Sullivan in touch with someone who could deliver cocaine.

"I just point the fingers here and we're there," he said. "That's all."

"But you set it up," Previte quickly added.

"That's all I'm doing," Luisi acknowledged.

Both agreed that sometimes you had to take a chance. "If you don't do that, you make no money," Luisi said.

"That's what I do, you know," said Previte. "What are we in it for? If I want to be safe, I'll get a job."

The cocaine was coming to him "on credit," Luisi explained. He would need the cash immediately. Previte said Sullivan already had the cash; he was just waiting for the delivery.

They walked and talked for several minutes, strolling around the North End, the predominantly Italian-American neighborhood where Luisi had grown up. Before coming back to the Caffe Vittoria, however, they returned to the topic of Borgesi and what he would or would not know.

"The other guy, George, you know how it is," Previte said.

"He told me straight out, 'No,'" said Luisi.

"The other guy above him says go ahead. He don't care."

"I know. No disrespect to George."

"No disrespect to nobody," said Previte.

Two days after the walk-and-talk outside the Caffe Vittoria, Luisi sent Bobby Carrozza to see Mike Sullivan at Irish International. The undercover agent and Carrozza, a loquacious thirty-nine-year-old wannabe wiseguy, spent more than an hour getting to know one another. The conversation was picked up on both audio- and videotape.

Sullivan said he had done some deals in Philadelphia but had been reluctant to try anything in Boston. He was still getting to know the city, he said, when he reconnected with Previte and learned of Philadelphia's interest in Beantown. Now he felt comfortable.

"Apparently . . . the Philadelphia side talked to the Boston side and everything was . . ."

"Copacetic," Carrozza replied.

". . . If I can make everybody some money and if I can make everybody happy and no one's pissed off at one another, why not take a shot? You know?"

"Right. As long as everything's all right."

Sullivan mentioned the diamonds-for-coke deal that hadn't worked out. Carrozza said he was aware of it, adding that "Everybody just missed each other."

But the phone call and the meeting at the café had gotten everything back on track, Sullivan said, steering the conversation in a direction that confirmed and reinforced all the acts in the drug conspiracy.

"I don't know if you realize, they talked to Joey that day," Sullivan said of the three-way phone call from his office.

"They did," Carrozza said. "I know that."

"The guy, he had needed his okay on it," Sullivan said of Luisi.

"Right," Carrozza replied.

Carrozza said the "thing" would be a "very good grade" and that he could deliver it later that afternoon. He also said he was told that, out of respect for Sullivan and his relationship with Previte, he shouldn't even bother counting the money when the transaction took place. Carrozza said he wouldn't have been comfortable counting out that much cash anyway. He just wanted the deal to go down smoothly. Sullivan said he felt the same way.

"If Joey Merlino and Bobby Luisi are talking on the phone and say it's gonna happen," Sullivan said, "I ain't about to fuck it up. You know what I mean?"

"Absolutely," Carrozza replied.

Two hours later, the young wiseguy returned to Sullivan's office carrying a briefcase. Inside were two neatly wrapped packages, each containing a white, powdery substance. Each package had the number 215 written on it.

Carrozza, not the brightest light, said he didn't know what that meant.

"That must be your code," he said to Sullivan. "Does two one five mean anything to you?"

The FBI agent explained that 215 was the telephone area code for Philadelphia.

The audio- and videotape captured it all. From the office, Carrozza and Sullivan walked over to the parking lot of the nearby Marriott Hotel, where another undercover agent—"He's family . . . my cousin," Sullivan had explained—was waiting with the cash. Carrozza was handed a satchel. He then walked over to a car that was being driven by Thomas Wilson, another Luisi associate. The pair drove away with fifty thousand dollars.

An FBI report indicated that the powder delivered that day weighed 2,093.73 grams—a little more than two kilograms—and tested out at 42 percent cocaine.

A week later, Previte met with Merlino. "The shit wasn't that good," he said, referring to the "two keys" that Carrozza had delivered. "He said he's gonna bring better stuff next time."

Ignoring the comment, Merlino mentioned that he had known Carrozza's father in jail. "He was my celly for two years, the father," Merlino said.

"Like I said, we got two from him and . . ." Previte said, bringing the conversation back to the cocaine transaction.

"How'd it go with the jewelry?" Merlino asked, changing the subject.

"Joey was smart," Previte said later. "The FBI kept telling me to get him to talk about cocaine, but he wasn't going for it. I couldn't force the issue, because then he would get suspicious. I just had to let it go. You can't make somebody say something. I couldn't use words like *coke* or *crack* or *cocaine*. I talked about the 'two keys.' What the fuck else would that be? I thought we had him pretty good with that, but they kept asking for more."

Previte handed Merlino twenty-five hundred dollars in cash at that meeting and said another twenty-five hundred was coming. It was his end of the two-kilogram deal.

Merlino took the cash and complained again about the IRS investigation. He also said he didn't trust most of the people around him.

"I swear to God," he told Previte. "Besides us, there's me, you, Georgie [Borgesi], Marty [Angelina]. I don't even want to fucking meet nobody. I swear to God."

Sullivan, on the other hand, was meeting most of the members of Luisi's crew in Boston and establishing a "comfort-level" arrangement in which they would all make some money. In June the undercover agent fronted another cocaine deal, agreeing to pay twenty-four thousand dollars in advance. He gave the cash directly to Luisi on June 2 at a meeting outside the Irish International office. The next day, Tommy Wilson, the driver in the first deal, delivered a package to Irish International. The meetings, cash transaction, and delivery were all picked up on tape.

Wilson boasted that he had "taken trains" in order to avoid being followed. "A little extra activity," he said, "but you know something, it's well worth it."

The powder Wilson delivered weighed 1,132.34 grams—more than one kilogram—and tested out at 51 percent pure cocaine.

At that point the FBI had three kilograms of cocaine, though not of the highest quality, and enough audio and video to make an airtight case against Luisi and his associates Carrozza, Wilson, and a fourth member of the crew,

Shawn Vetere. But the feds wanted more. They wanted to push the deals over the five-kilogram level in order to enhance the penalties. Under federal sentencing guidelines, deals of five kilograms or more carry a mandatory sentence of ten years to life.

Over the next three weeks, Sullivan, Luisi, and Previte would discuss another major transaction. Sullivan, continuing his underworld seduction, quietly slipped Luisi five hundred dollars in cash during one meeting. It was just between the two of them, he said, a way to thank him for everything he had done. Luisi, still wearing his Rolex, smiled, pocketed the cash, and said he would do whatever he could to make the next deal happen.

Sullivan wanted "three or four" keys. At twenty-four thousand dollars a kilo, Luisi said he could put it together, but he'd need Sullivan again to front the money. At that point the FBI balked.

"They didn't want to put up the cash," Previte said. "They wanted the deal—the three or four kilos would have gotten us over the five-kilogram mark—but they didn't want to put the money out first. And Luisi didn't have the cash to buy from his supplier unless Sullivan fronted him. What a fucking outfit this was.

"Meanwhile, the U.S. Attorney's Office gets into the act. Now they're saying they want more tape of Merlino. They're not sure we got enough. I know we do, but I tell them I'll try to get more. And Joey's more cautious than ever."

On June 8, Previte met with Merlino, handed him another twenty-five-hundred-dollar payment, and told him, "I'm going to Boston tomorrow and we'll get three more and then we'll have ten G's to split."

Still guarded with his words and worried about a drug arrest, Merlino pointed to his nose and said, "That's good, but be careful up there with Bobby. That's a bad pinch."

That same day, Sullivan met with Carrozza to talk about the details of the pending transaction. Carrozza boasted that everybody—Merlino, Luisi, Vetere—was pleased with what was going down. "Everybody's happy with this," he said. "They want to make it work."

The last tape Ron Previte made during his two-and-a-half-year undercover operation was on June 10, 1999. It was a conversation with Joey Merlino, and in many ways it captured everything the young wiseguy was about.

Merlino had attended a Phillies game the night before, and Previte said that he'd heard him mentioned on WIP, the all-sports talk radio station in Philadelphia, that morning.

The Phillies had played the Yankees. George Steinbrenner, the Yankees' cantankerous owner, was sitting in a box seat right behind his team's dugout with Philadelphia mayor Ed Rendell. Merlino said he was sitting right behind them. A friend, he said, "gave me two [tickets]. The best seats in the house. Right on the field."

Previte handed Merlino one thousand dollars and told him the deal in Boston was on hold, that it had hit a snag, that Luisi was holding things up by insisting that Sullivan front all the money.

"You gotta do me a favor, you gotta talk to him," Previte said. "We got a thing goin' with them this week, Joey, I'm tellin' ya. . . ."

"Did ya ever ask, did ya ever ask the guy about the twelve watches?" Merlino said, referring to an earlier discussion in which Sullivan had told Luisi he could get his hands on a dozen Rolex watches.

That deal had fallen through, Previte said, and tried to swing the talk back to cocaine. "Ya gotta do this for me," he told Merlino.

"All right, I'll do it," Merlino replied. "But tell your guy when he gets the watches, I'll sell them on the street myself. . . . I can sell them in two days. . . . That's better than fuckin' money . . . the Rolex. . . ."

"Will you listen to me one minute, please," Previte said in

frustration. "I got a twenty-five, fifty-G deal goin' this week. I don't want no bad shit."

"All right."

Previte said Sullivan's source for the hot watches had "run off with them . . . Fuck the Rolexes," he said.

But Merlino was obsessed with the idea of moving hot watches. "If he gets, listen . . . I can sell them. I, listen, I'll sell them in two days. They'll be gone. . . . Tell your guy to Federal Express them to your house or whatever. . . . I got ten kids that want them. And they're, they're better if they're the ones with the two tone, you know what I mean?"

"Understood," Previte said. But he was still pleading that Merlino talk with Luisi. "Just tell him, 'Do the right thing.'"

Merlino said he would. "I'm gonna say, 'Listen. What Ronnie's got up here. You gotta promise me. You gotta do the right thing. Treat him like he's . . . one of us.'"

Previte then asked Merlino to tell Luisi "don't jump on it no more."

"You got it," Merlino said. "You got my word."

The conversation then swung back to the Phillies game and the morning radio talk show. "They said on the radio I was at the game?" Merlino asked Previte.

"Yeah, some guy called in and then the guy, what's the Italian guy in the morning . . . Cataldi [Angelo Cataldi, a popular morning show host]. They said there was Steinbrenner, the mayor, and Joey Merlino with two blondes. So I hope your wife don't hear it."

"Blondes," Merlino said. "Motherfuckers. I had, I was with fuckin' Frank Gambino."

"That's what they said," Previte said, laughing as the conversation ended.

For the next two weeks, the feds danced with Luisi over the cocaine deal. It was on, then off, then on again. Ultimately the feds decided to pull the plug. There would be no more cash up front. There would be no more tapes. Instead, there were arrest warrants.

They came for Joey Merlino early on a Monday morning, when they knew he would be sleeping off a weekend of carousing in the bars and restaurants around Margate.

At 7:10 A.M. on June 28, 1999, two members of the FBI's Organized Crime Squad, Special Agent Jack Meighan and Philadelphia Police Detective Marc Pinero, knocked on the door of apartment number 8 at 25 South Adams Avenue. Meighan had his gun drawn.

What happened next, like so much else in the life and times of Skinny Joey Merlino, is open to interpretation. What is clear is that the face of the Philadelphia mob changed forever that morning at the Jersey shore.

Merlino's wife, Debbie, heard the knock at the door. She had just gotten out of bed to see about her eighteen-month-old daughter, Sophia. The Merlinos' first child, Nicolette, who was about to celebrate her third birthday, was asleep in another room. So was Joey's mother, Rita, who had spent that weekend at the shore with them.

Debbie Merlino put the baby in bed with her husband and went to see who was knocking. "I heard this pounding on the door," she later testified. "I opened it and saw a gun pointing in my direction."

The two men standing in front of her told her they were there to arrest her husband. She told them he was asleep upstairs. With that, she said, they brushed past her and headed toward the second-floor bedroom.

Debbie Merlino panicked, screamed out, "The baby's in bed with him!" and raced after the men with the guns. She burst into the bedroom on the heels of the two arresting officers, scooped up Sophia, and took her into the adjacent bedroom where Nicolette was sleeping.

Meighan holstered his gun after spotting the child in bed.

He and Pinero told Merlino they were there to arrest him for conspiracy to distribute cocaine.

Merlino got out of bed. He asked to use the bathroom.

Meighan and Pinero noticed a pair of men's blue jeans on the floor. Meighan picked up the pants. They were heavy. He checked the pockets. Inside, he found two wads of cash, each wrapped in a rubber band. Later, when they counted the money, they discovered that one wad contained ninety-six hundred-dollar bills. The other contained five hundred and eighty dollars in different denominations. Merlino's driver's license was with the cash.

"Those are not my pants," he said at first. "That's not my money."

But when he got dressed, he slipped easily into the jeans.

"They seemed to fit perfectly well," Meighan said.

They cuffed Skinny Joey and brought him down the stairs. Rita Merlino, who had been through this kind of thing before, demanded to see a warrant. They tried to ignore her. Short and stocky, with the mouth of a South Philadelphia truck driver, Rita Merlino is not easily ignored.

There was shouting. Profanity. Shoving. The scene spilled out into the street. People staying in the condos nearby began to peer out their windows and stare from their porches.

Rita Merlino, causing what Pinero would later describe as an "uproar," kept demanding to see the warrant as Pinero led her son in handcuffs out of the house and to the backseat of the unmarked car the feds were using that morning.

She was still screaming when Marty Angelina, who was renting a summer home nearby, came running up. Like Rita Merlino, he became excited and agitated. Angelina, who stood about five foot six and weighed close to two hundred and fifty pounds, was flailing his arms "like a drunken penguin," Meighan later reported.

He asked for and was shown a copy of the warrant. Then

he refused to give it back. There was a brief scuffle. He was pushed into some bushes. The warrant was ripped out of his hands by Meighan. Pinero, meanwhile, was trying to restrain the even more vociferous Rita Merlino.

And from the car, in handcuffs, Joey Merlino screamed, "Get your fuckin' hands off my mother, you fuckin' Puerto Rican bastard."

It was a South Philadelphia street scene, played out against the backdrop of a sun-dappled Margate morning.

At the very same moment Merlino was being arrested, five associates of Natale's were being scooped up at their homes in South Jersey and Philadelphia. Anthony Viesti, Ray Rubeo, Johnny Santilli, and Robert and Thomas Constantine were charged that morning with conspiracy to distribute methamphetamine.

Natale, already in jail on the parole violation charge, was informed that he too was being charged in the meth case. Lodged in a federal prison in Ohio at the time, he was told that he would be sent back to Camden, New Jersey, for a formal arraignment.

That same morning in Massachusetts, FBI agents knocked on the doors of Robert Luisi, Shawn Vettere, Bobby Carrozza Jr., and Tommy Wilson. All were charged with conspiracy to distribute cocaine.

As the arrests took place, Ron Previte sat in a "safe house" about sixty miles from Philadelphia. He had been spirited out of Hammonton the previous Friday morning and had spent a long and boring weekend in protective custody.

It was a new experience, he said during a brief telephone conversation that weekend. "Friday night, nine-thirty, they're brushing their teeth and puttin' their jammies on," he said of his handlers. "We go out to dinner the next night, and when the bill comes, they start calculating who owes what. Down to the penny. They think the Ground Round is a great restaurant. I can't live like this."

Then Previte turned philosophical. He had driven around Hammonton on Thursday afternoon, he said, looking at familiar places and waving to friends he had known for years. He wasn't sure when he would see them again, or how they might look at him in the future.

"I'm looking at people . . ." he said, his voice trailing off.

"Compassion? I don't know. I had to jump when I jumped. In a way it's like dying. My life as I know it is over. It's finished."

Then he laughed. "I had some clothes at the dry cleaner. The guy said, 'They'll be ready on Thursday.' I said to myself, 'They better be, 'cause I can't pick them up on Friday.'"

On the Monday morning when the arrests went down, Previte listened to the radio, watched the news on television, and talked via phone several times with the FBI agent handling his case and with a reporter who was working the story.

He didn't know what to expect, he said. He just knew his life was about to change, and he began to wonder if that change was for the better.

There are times, five years later, when he still wonders.

16

Ralph Natale said it hit him all of a sudden as he sat at the defense table in the federal courthouse in Camden, waiting for his arraignment in the drug conspiracy case.

It was Monday morning, July 19, 1999. Natale, looking tan and fit, had been brought back from the federal prison in Elkton, Ohio, where he was serving time on the parole violation charge.

The hearing, in a third-floor courtroom of the modern new federal courthouse in downtown Camden, lasted about five minutes. Natale walked confidently into the courtroom dressed in a white T-shirt and tan prison-issue pants. He smiled and blew a kiss to his wife, Lucia. Then he spotted FBI Agent James Maher, the supervisor of the Organized Crime Unit for the Philadelphia office.

Maher, a veteran investigator who was soon to retire, was going to cap his career with the Natale and Merlino cases. He had been tracking the Philadelphia mob for more than twenty years, beginning in the Bruno era. He had been one of the agents involved in the early drug and arson cases that sent Natale to prison in 1979. He had worked Scarfo. He had worked Stanfa. Now he was working Merlino and Natale.

"Mister Maher," Natale said with a smile, "I'm tired of looking at you in courtrooms. Give it up."

Maher grinned and shook his head.

Natale pleaded not guilty that morning, but he had already begun to consider his options. When he looked at his wife and several other family members who had shown up in court that day, he would later explain, he decided that he, not Maher, would be the one giving it up.

"I gave my whole life to La Cosa Nostra," Natale said. "My whole adult life. I did almost sixteen years in prison for them. I came out. I did what I had to do and I went back in prison. The day they brought me in from Ohio . . . I looked at my family's faces, and I found out what I truly did to them because of what I thought and who I was all my life."

Natale would tell that same story several times over the next three years as he appeared on the witness stand for various legal proceedings. It always came out the same way: this was his epiphany. La Cosa Nostra, he said, was a "descent into Hell. . . . I was there . . . at one level or another for almost forty years. . . . I lived it in my heart and in my soul." That day, for the first time, he said, he saw what it had done to his family. And that day, he decided "no more."

"If there is any life left for me," he said, "I'll give it to [my] family. No more La Cosa Nostra."

No one bothered to ask if the prospect of spending the rest of his life in jail might also have had an effect on his decision. No one asked if he was thinking about his family when he came out of jail in 1994 and went right back to dealing drugs, or if his family was a priority when he quickly grabbed the top spot in one of the most dysfunctional and wantonly violent Mafia organizations in America. And no one asked if his beloved wife and loving children were on his mind when he took up with a twenty-five-year-old who was once his daughter's best friend.

It was better, at least for the feds, not to ask.

Natale's sales pitch was built around his epiphany. That and the promise that he could put Joey Merlino and several of his top associates in the middle of half a dozen murder conspiracies. That was what Natale said he could bring to the table.

For the FBI it was a no-brainer. Here was a chance to send

Skinny Joey away for the rest of his life. Maybe get him sentenced to death. And it would be accomplished by flipping a sitting mob boss—an unprecedented development in the underworld, and a move that would enhance the national reputation of the Philadelphia FBI. There had been acting bosses who had cooperated, like Alphonse D'Arco of the Lucchese organization. And there had been former bosses, like Angelo Lonardo out of Cleveland, who had struck deals after they were sentenced to heavy jail time.

But this was a boss who was still running the family, albeit from prison. A boss who over the past four years had, in theory at least, been involved in every major decision—who supposedly knew about the murder plots, the extortion attempts, the loan-sharking and gambling operations.

He was, of course, a twice-convicted narcotics trafficker who had been nailed again dealing drugs. And he would eventually admit to his own involvement in eight gangland murders—two by his own hand—and four murder attempts.

Was the case against Merlino, and the evidence stacking up against his associates, strong enough without Natale?

Was society better served with both Natale and Merlino doing hard time, or was it better to cut Natale a break in the hope of ensuring the conviction of Merlino?

Does it make sense to make a deal with the boss in order to nail the underboss?

Those were some of the issues that surfaced once Natale raised his hand and said he wanted to cooperate.

The fact that such questions were even being considered spoke volumes about the state of the Philadelphia mob and the attitude of those investigating and prosecuting it. Mob informants had been an effective tool in Philadelphia for fifteen years. The demise of the America Mafia could be traced to the testimony of Nick Caramandi and Tommy DelGiorno back in the late 1980s. They, in turn, begat Phil Leonetti, the Scarfo underboss who was one of the most effective and devastating government witnesses in underworld history.

Leonetti in turn begat Sammy Gravano, if not the best then certainly the most celebrated American Mafia informant.

But where do you draw the line in making a deal with the devil?

Would the feds in New York have negotiated with John Gotti to get Sammy Gravano? Unlikely.

Natale was offering Merlino, his underboss. But he was also offering the city's celebrity gangster, the "young punk" who dined at the finest restaurants, who signed autographs at sporting events, who thumbed his nose at the FBI when he hosted Christmas parties for homeless kids, when he gave away turkeys in the projects, when he celebrated his daughter's baptism in a fancy ballroom with four hundred guests.

It wasn't about justice. It was about getting Joey.

The feds took the deal.

Ralph Natale had finally talked his way into a score.

The negotiations began shortly after Natale entered his not guilty plea. His wife made the first call to Maher. Secret meetings followed, and before long the rumors were spreading. By mid-August, the media were reporting that a deal was in the works—that Natale was going to cooperate.

One story in particular incensed the jailed Mafia don. Under the headline "The Mob, the Moll and the Boss," the story, based on "sources," outlined Natale's fractured relationship with Merlino, and pointed to his affair with Ruthann as part of the reason for his decision to turn government informant.

Natale was convinced that those in Merlino's camp—particularly George Borgesi—had leaked the information in an attempt to further discredit him. But stories about Natale's affair, and how it was undermining his dealings with the younger members of his mob, had been circulating widely for more than a year. Previte had been shaking his head in disbelief for months while talking about Natale and his "gun moll." Law enforcement investigators in New Jersey and

Philadelphia frequently joked about the soap opera that was playing out behind the scenes of the broader investigation. It was a story most reporters thought would never be told. It had no relevance until Natale decided to cooperate. Then it became part of the explanation for why he was turning on his former crime family members.

Ironically, Natale would later confirm that fact in phone conversations from prison that were routinely taped by authorities. In discussions with his wife and with Danny D'Ambrosia, Natale complained about how he had been abandoned by the Merlino group. From the witness stand he expanded on the subject, complaining that Merlino and the others had stopped sending money to Ruthann and to his wife. Ruthann, he said, was supposed to get the one thousand dollars that Tony Machines Milicia was paying to Natale each month. Instead, after one payment the money stopped. His wife was also supposed to get monthly support from his end of the other mob gambits. She too was cut off.

"I thought Joey Merlino and the rest of the fellows would take care of my family," Natale said. Instead, he said, "it was like I had died."

In one phone conversation with D'Ambrosia he referred to Merlino as a "little punk" who thought he was a "combination of Nicky Scarfo and John Gotti." He described Borgesi as a wife-beater and "half a homo" who had orchestrated the media "propaganda" against him. Mazzone, he said, was strung out on cocaine; Joe Ligambi, the veteran mob soldier who became acting boss after Merlino was arrested, was simply "a freak."

Natale formally signed his cooperating agreement on September 11, 1999, a date that would attain national significance two years later. Then it simply marked the beginning of the end of another era in the Philadelphia underworld.

The war of words continued for most of that fall. Previte's role as an informant surfaced as soon as Natale and

Merlino were charged in the drug cases, and Merlino's associates, especially Borgesi, badmouthed Natale and Previte continually. No longer "Mighty Joe Young," he was now "the Fat Rat."

The media loved it. Skinny Joey and the Fat Rat. It had a certain ring.

But most of the venom was aimed at Natale, perhaps because of who he was or who he claimed to be. Perhaps because of what he might be able to do.

"He used to talk about how he was a 'man's man,'" Borgesi said bitterly one day after it was clear that Natale had cut a deal with the feds. "He used to brag about how he would look in the mirror each morning and see 'the boss,' the only person who could tell him what to do.

"What's he see in the mirror now?"

Natale was shifted to an undisclosed cooperating witness wing of the federal prison shortly after he began cooperating. From there, by phone, he continued to communicate with his family and members of his "former family." Danny D'Ambrosia became a middleman, transferring messages between Natale and Pete the Crumb Caprio up in Newark. Natale said he was not going to "hurt" Caprio, but wanted him to consider cooperating. Natale also predicted that he would be home "soon." In some conversations it sounded as if he thought he'd be in jail for less than a year. Most of the conversations were cryptic, however, and could be interpreted in several ways.

In one, he predicted that Merlino's associates would have a big surprise by Christmas. Most people later interpreted that to be a reference—accurate, as it turned out—to a superseding indictment that would expand the charges against Skinny Joey and add new defendants to the case. But there were other events that transpired during the fall of 1999 that raised questions about just what Natale was trying to do while in federal custody.

Caprio was the key.

Pete the Crumb became an ambitious man in his old age. With Natale, Merlino, and so many others falling by the wayside, he decided that he, rather than Joe Ligambi, should be the boss of the Philadelphia mob. As he later admitted to federal authorities, even as Natale was sending him messages through D'Ambrosia, he began plotting with members of the Gambino and Genovese families to take over the top spot in the Philadelphia mob.

The plan, he said, was to lure Borgesi, Ligambi, and Mazzone to a meeting in North Jersey, where they would be killed. "We were gonna whack 'em all out," Caprio later testified. The plan was to take the bodies to a construction site where they would be buried. It was strictly business, said the then seventy-one-year-old gangster. "I had no ill will," he said of Borgesi, Ligambi, and Mazzone. "This was the way the mob works. . . . It wasn't personal."

Members of the Gambino and Genovese families, he said, were ready to recognize him as boss. They did not recognize Ligambi. What they wanted in exchange, he said, was the freedom to develop their own gambling and video poker machine operations in Philadelphia.

As always, it was about the money.

What Pete the Crumb didn't realize was that his triggerman, the former child molester Philly Fay Casale, had already cut a deal with the government. Casale had been wearing a wire during several meetings with Caprio in the fall of 1999, recording Caprio as he implicated himself in the Sodano murder.

Even as the Philadelphia mob was coming apart at the seams, it remained one of the most recorded crime families in history. Natale's prison phone conversations, Casale's body wire, and a Pennsylvania State Police wiretapping operation aimed at Borgesi's bookmaking business provided additional hours of mob talk.

One conversation, from a bug placed on a phone at Pasta,

Cheese & Things, the South Philadelphia food distribution business run by Borgesi and Fat Angelo Lutz, would later prove devastating for Borgesi. While other tapes detailed his disdain for drugs, offering his lawyers occasion to paint him in a positive light to a jury, the state police tape offered a chilling look at a wiseguy who was both vicious and mean-spirited. The tape was Borgesi in his own words—Borgesi, prosecutors would argue, as he really was.

An athletic five foot eight with a barrel chest and a legendarily short temper, Borgesi had always belittled Lutz, the short, squat mob associate who played Santa Claus and whooped it up at the Mummers Parade. In November 1999, with tension building about the ongoing racketeering investigation, Borgesi learned that Lutz—not for the first time—had tried to cash a two-thousand-dollar check drawn on the pasta business without his permission. Then, when Borgesi confronted him, Lutz lied about it. Borgesi, talking on the phone with a bookmaker friend of his, recounted what happened next.

"I knock 'im out, actually knock 'im out," Borgesi said, laughing into the phone. "He was sleeping, snoring on the ground . . . for ten minutes. I hadda smack him to wake him up. I fuckin' punched him. He went down and I kick 'im right in the face. . . . He just went right out. He slid down the wall like in slow motion."

Borgesi said he didn't intend to "crack" Fat Angelo. "I just said, ya know, 'Fuckin' jerkoff, I should break your fuckin' head, you no good motherfucker. Fuck you.' And then he like answered me fresh, and then I went fuckin' berserk. I tell ya, if I had a gun, I woulda shot him. I was lookin' for a gun too. I was fuckin', like I went nuts. Yo, you woulda died. I kicked 'im so fuckin' hard he got knocked right the fuck out. And he was actually sleeping. I was going berserk. They had to stop me 'cause I woulda killed 'im. He was out and I was still kicking 'im.

"You should see his nose. His nose gotta be broken. It's

all distorted and all. His nose is bleeding. His mouth was bleeding. Then I'm screaming like for five minutes. I had ta walk outta the place 'cause I was screamin', I was gettin' a headache from screaming.

"Then my brother was saying . . . 'He ain't movin'. I think he's dead. He ain't movin'. . . . I smacked him in the face to wake 'im up.'"

"Where is he now?" asked the bookmaker.

"He's at the place," Borgesi said. "Call 'im up. Say, 'Yo, Ange, what're you doin'?' He'll say, 'Nuttin'.' Say, 'What's the matter? You sound like you got a headache or somethin'.' I'll call you right back. Break his balls, all right?"

Borgesi wasn't the only one with a short fuse and a big mouth.

Merlino found himself jackpotted shortly after his arrest in June in the cocaine case. Within weeks, he was also charged with threatening the life of a federal officer. While in the car with agents Marc Pinero and Jack Meighan, Skinny Joey apparently went on a rant, aiming most of his bile at Pinero, a Philadelphia police officer who had been assigned to the FBI Organized Crime Task Force for years—and who'd known Merlino since he was a teenager.

Among other things, Merlino called Pinero a "fucking FBI wannabe" and a "fucking jerkoff like your brother agent . . . the one who got his face shot off down on Delaware Avenue." This, Pinero knew, was a reference to Chuck Reed, a highly regarded veteran FBI agent who had been killed during an undercover drug bust gone bad.

Merlino also berated Pinero for arresting him in front of his wife and kids. Then, lowering his voice, he said coldly, "You have monkeys at home too. What goes around, comes around."

Pinero said he interpreted that as a direct threat against his family.

For a week in October, a federal jury in Camden heard all

the details about the day Merlino was arrested, including the altercation outside the house, in which Rita Merlino and Marty Angelina confronted the arresting officers, and the exchanges in the car, in which Merlino allegedly threatened Pinero.

On most days, the courtroom was packed. Merlino always drew a crowd. In addition to curiosity seekers and the media, the trial was a daily mob scene. Borgesi, Mazzone, Angelina, even Ligambi—whose disdain for authority was matched only by his aversion to reporters and television cameras—showed up. And while prosecutors were prohibited from making any reference to Merlino's role in organized crime—it was deemed irrelevant to the case at hand—the presence of the brooding mobsters, a crew right out of central casting, clearly let the jury know what it was dealing with. Assuming, that is, that any members of the panel had somehow not heard of Skinny Joey Merlino before they were summoned to jury duty.

Pinero testified that he took the threat seriously because of the tone and nature of what Merlino said. Meighan also testified, backing up his partner. Deborah and Rita Merlino told the jury about the morning of the arrest and the tension inside and outside the condo.

The trial lasted a week. The jury deliberated for less than three hours.

Merlino was found not guilty. The wiseguys in the back row of the courtroom applauded. Deborah Merlino said she was grateful. Skinny Joey smiled and waved as he was led out in handcuffs. He was still being held without bail in the drug case.

It was a victory for the mob, but clearly a minor skirmish that was part of a bigger war. Merlino's lawyer, Edwin Jacobs Jr., might have described it best in his opening argument to the jury. The government, he said, was "making a federal case" out of what had simply been an angry confrontation, brought on in part by his client's rage and embarrassment at being arrested in front of his wife, mother, and two young daughters.

"An insult is not a crime," Jacobs argued. "Profanity is not a crime. . . . Unless you happen to be Joey Merlino."

The jury verdict came in on Thursday, October 21, and was the headline story in the papers the next day. That Friday, Ron Turchi, who had served briefly as Natale's consigliere, left his home in the 1100 block of South Seventh Street in South Philadelphia to meet someone for lunch.

No one who cared about him would ever see him again.

The next day, his wife reported him missing. Three days later, police found the black 1992 Toyota that Turchi was believed to be driving parked in the 900 block of East Passyunk Avenue, a couple of blocks from his home. They towed the car to the police pound and, after obtaining a search warrant, popped the trunk.

Inside, naked with a plastic bag over his head and with his hands tied behind his back, was what was left of Ron Turchi. The sixty-one-year-old wiseguy, who for years had straddled the fence during the factional and often deadly squabbling that was the mark of the Philadelphia mob, had been beaten and shot to death.

Investigators believed that the car was left on a public street so that it could be found. The murder, they said, was clearly a message.

No one has ever been charged with the Turchi slaying, but several months later, after he began cooperating, Caprio said that Ligambi had told him, "We dimed Turchi out to send Ralph a message."

The fact that Turchi had fallen out of favor with Natale, and had been demoted from consigliere to soldier, apparently was immaterial to those who killed him.

Previte, like the FBI, could only speculate. But like the FBI, he knew the signs. "It was brazen, in-your-face," he said. "It's the way Nicky Scarfo used to do things."

Turchi had been grabbed within twenty-four hours of

Merlino's acquittal in the Pinero case. It was, the feds believed, another example of the arrogance of the local mob.

One other wiseguy had his own thoughts about the timing of the murder.

Turchi's body was discovered on October 26. That was the birthday of Gaetano "Tommy Horsehead" Scafidi. He was supposed to be getting out of prison early in 2000. And he was on a lot of people's minds.

The feds had begun visiting him, trying to convince him to cooperate. Their message was clear and to the point: if he didn't agree to work with them, he would probably be indicted again. So even if he got home, his stay would be short-lived. There was a RICO case coming up against Merlino and the others. He could be a defendant or he could be a witness.

There was, of course, one other option, and Tommy was well aware of it. Merlino and those around him would kill him for switching sides to join Stanfa in the middle of the bloody 1993 war.

When Scafidi heard about the Turchi hit, he took it personally. "He had a plastic bag over his head and he was stripped down naked, found naked in his car," Scafidi later told a jury when discussing the Turchi murder and its implications. "And I knew that was a sign that they stripped him down to see if he was wearing a wire. That's what I believe because that's the way the mob works. They stripped him down to see if he had any kind of listening devices or tape recorders on. And I knew that was a message to me to keep my mouth shut, because they found this guy on . . . my birthday."

Tommy Scafidi turned thirty-five that day. He was scheduled to be released from prison in six months. He wanted to live to be thirty-six.

On December 16, the feds announced that the indictment against Merlino was being expanded to include his involvement in the stolen property ring. Four others, including Ralphie Head Abbruzzi and Frank Gambino, were also charged that day. In a separate indictment, nine members of the stolen property ring were also charged.

The joint indictments outlined the operation, in which over $1.3 million in goods, ranging from sweat suits and VCRs to toy trains and bicycles, had been stolen from terminals along the Delaware River. The mob, the indictment charged, had extorted its way into the deal, siphoning off stolen property worth over four hundred thousand dollars.

Previte's tapes were a key to the case. But so was the testimony of two members of the stolen property ring who were also cooperating. One was the hapless hijacker who was discovered trying to record Gambino during the card game.

At a press conference announcing the new charges, Michael Stiles, the U. S. attorney for the Eastern District of Pennsylvania, told reporters, "We haven't stopped. We are continuing. Other charges could follow."

It was clear that nothing in the indictment came from Natale. Whatever he was giving the government was apparently still being processed.

With Merlino in jail for Christmas that year, there was no high-profile party for the homeless, no turkey giveaway in the projects. For the rest of the mob it was a quiet holiday season, a time to wait and wonder.

"These guys all know what's coming," Previte said, "because they know what they did and they know who they did it with. I'm sure they're all sitting around wondering, trying to remember what they said to me and asking themselves if I was wearing a wire when they said it. They all got problems. It's just a matter of time. It's not if, but when."

Previte, even without an indictment hanging over his head, was just as restless.

He had left protective custody now and was back on his own, roaming around South Jersey and occasionally popping up in Philadelphia. He would still visit a barber in his old West Philadelphia neighborhood. And there would be reports of "sightings" of the Fat Rat in Hammonton and Atlantic City.

The FBI was concerned, but knew they couldn't force him to do anything. He hadn't been charged or arrested. (He would later be indicted for his role in the DeLaurentis case, but was immediately released on bail.)

Even after the Turchi murder, Previte stayed out on the street. He is still there today.

"I couldn't stay with the FBI," he said. "It was terrible. I told them, 'I don't care if I get shot. I can't live like this.' I just couldn't hang with those guys. I knew I had to get the fuck out of there."

Within a month, he had left the safe house. For a time he lived in a Ramada Inn in New Jersey, not far from his roots. Eventually, and amazingly, he bought a house in a rural area and moved in, using a different name, but changing hardly anything else about his appearance and lifestyle. He has since moved once more.

"For a while I was a vagabond," he said of those days. "But I couldn't stay with the FBI. It was just something I couldn't do. I had to be on my own. After I left I had to call in all the time, and they would get nervous sometimes if they didn't hear from me, but I wasn't worried. I told them, 'This mob ain't shit. Who the hell they gonna kill?'

"I went about my business. Sometimes the best defense is to stay on offense. It's good when people can't figure you out. Believe me, I was armed and dangerous, but I never had any problems."

17

"And what did you conclude about your future in the mob?" the prosecutor asked.

"I didn't have one anymore," Tommy Scafidi said.

"How did you feel about that?"

"Sick to my stomach. I kept my mouth shut all them years. It didn't mean nothing."

Tommy Horsehead said his lawyer came to see him on New Year's Eve, 1999, and asked him to think about making a deal with the government. The lawyer, Chris Furlong, told Tommy it was his decision to make, that he would do whatever Tommy wanted. But he also told him that he was convinced Scafidi would be killed if he returned to South Philadelphia after he was released from prison.

"I was just hoping if I kept my mouth shut, I would get a pass," Scafidi said. But in his heart, he said, he knew that wasn't possible. He knew Merlino and those around him too well.

Merlino's plan all along, Ralph Natale later told authorities, was to give Scafidi a false sense of security, get him back to South Philadelphia, and then kill him. The fact that Merlino was behind bars didn't change any of that.

"Merlino felt that he could lure Scafidi in by telling him all was forgiven," Natale said.

Two weeks after his New Year's Eve visit, Scafidi called a family member and arranged to get a message to Jim Maher, the head of the FBI's Organized Crime squad—the same agent Natale contacted when he began negotiating his deal. On February 1, 2000, two FBI agents visited Scafidi at the federal prison in upstate Pennsylvania where he was being held. Scafidi was to be released in two months. During the meeting Scafidi told the agents he wanted to cooperate. He told them what he had done. And he also made a point of telling them what he hadn't done.

The feds, for example, believed he had been involved in the ambush of John Stanfa and his son on the Schuylkill Expressway in August 1993. It was one of several pieces of information that Natale had gotten wrong. Scafidi had already broken with Merlino at that time, he said; he'd had nothing to do with that hit. His name had also surfaced in connection with the 1990 slaying of Louis "Louie Irish" DeLuca, a South Philadelphia bookmaker. Again, Natale had been the source for that information; again, Scafidi said he'd had nothing to do with it.

To protect himself in jail, Scafidi cut his first meeting with the FBI short and informed the agents that he was going to create a scene as he left the room. He then got up, headed out the door, and in the hallway—where he knew other inmates would hear him—he screamed, "Don't you fuckin' come up here no more. . . . I got nuthin' to say."

In a matter of days, Scafidi was transferred to the Bucks County Prison in Doylestown, a short thirty-minute car ride from Philadelphia. There he was able to meet regularly with the FBI and begin to negotiate his cooperating agreement.

"I cried like a baby before I did this," Scafidi would later tell a jury. "I mean, I know what it is to be a stand-up guy and have people rat on your family members and everything. . . . This wasn't an easy thing for me to do."

Scafidi, a fourth-generation South Philadelphia wiseguy, said he had tried to find another way. He had reached out for his brother, Salvatore (Tory), who was still in prison. He had also made contact with some made guys in New York, whom he had met while in jail.

But the word came back that there was nothing anyone could do. No one could help him.

"If they would have, I don't know, if they would have sent a message saying 'You would have got a pass,' or some-thing . . . They didn't give a shit about me. . . . The only people I could go to for help was the FBI."

By the time his former associates did reach out for him, Scafidi said, it didn't matter. He didn't believe what he was told.

From the time he went to prison in 1994 through the fall of 1999, "when Ralph went bad," no one in the organization had made any effort to talk with him. No one did anything for his family—his wife, Nella, and their young son. No one cared.

But shortly after Natale began cooperating, Scafidi said he started to hear, indirectly, from the guys he had grown up with. A lawyer who came to see him in the fall of 1999 told him not to believe everything he was hearing, that he didn't have a problem on the streets. The lawyer said he had that directly from Merlino.

After word got out that he'd been transferred to Bucks County—a move that immediately raised suspicions among defense attorneys and wiseguys alike—Scafidi heard even more.

Joe Santaguida, one of Merlino's lawyers, visited him at the Doylestown jail in February and told him both Joey and George Borgesi sent their regards. This, Scafidi noted, came nearly six years after he first went away. Now his old boy-hood friends were asking about him.

Santaguida also told him that Borgesi was a "nervous wreck," that he was in his law office every day asking what was going on, speculating on who was going to be indicted, trying to figure out what the feds had and who was cooperating.

Santaguida said Borgesi wanted him to know that "every-thing was all right . . . that Ralph doesn't know anything firsthand . . . that Ralph can't hurt anybody."

Without admitting that he was cooperating, Scafidi told Santaguida that he didn't understand this sudden concern. "All of a sudden youse are worrying about me," he said. "When I tried to get in touch with youse in 1996 . . . nobody gave a shit about me. Now that their boss went bad, everybody's worrying about me."

Again, Scafidi said, he was assured that Natale had no firsthand information and that the feds would never bring murder or attempted murder charges unless they had somebody who was at the scene of the crime. That description fit Scafidi, who said he was there, with a gun in his hand, on the day Joey Chang got hit. He was exactly the kind of witness the feds needed to corroborate Natale. But he wasn't about to share that information with Merlino's lawyer. Instead, he told Santaguida that the FBI "was breaking them for me." They were "busting my balls," he said, coming up to see him and telling him that he was going to be indicted and that if he went home he was going to be murdered.

"I did my time like a man," Scafidi said. "And I'm still hearing from people that I'm going to get killed when I come home. And I did the right thing all my life."

Scafidi's prison meeting with Santaguida lasted about an hour. Among other things, the lawyer told him that Borgesi wanted to send a hundred dollars to his prison commissary. Scafidi said no, thank you.

A few weeks later, he got a letter from Borgesi in which the mob family's new consigliere went on at length about old times and about how things had gotten screwed up. The bottom line was: Tommy Horsehead could come home, he would have no problem with La Cosa Nostra. It was the first time in six years that Scafidi had heard from his old friend.

The last time they had spoken was back in October 1993, shortly before Scafidi had switched sides and joined the Stanfa organization. Scafidi had met with Borgesi at the Thirtieth Street train terminal in Philadelphia. Tommy

Horsehead had picked the massive and very public train station because, even back then, he feared that he might be targeted. They met in the middle of the sprawling, cathedral-like building with throngs of travelers coming and going. Scafidi had several friends stationed around the hall, on the lookout for possible hit men. He and Borgesi talked for about twenty minutes that day. Without telling him directly, Scafidi was effectively explaining why he planned to switch sides. Merlino, he complained, had stolen money from everyone. Scafidi estimated that he was owed between two and three hundred thousand dollars, money he had collected during the shakedown of bookmakers and gamblers in 1992 and "turned in" to Merlino and Mike Ciancaglini—money that Merlino had promised would be divided up among the group at Christmastime. Scafidi never saw any of the cash. Merlino, he said, was selfish and treacherous.

"Joey is for Joey," Scafidi told Borgesi that afternoon. He said that he knew Merlino wanted to kill him.

"Joey screwed everybody," Scafidi later told a jury while explaining in detail the meaning of Borgesi's letter. The entire time he had been in prison, Scafidi said, "I never got an envelope for Christmas. My family never got a hello. All of a sudden they're worried about me."

That was clear from the letter Borgesi had sent.

"A lot of the things you said to Joe [Santaguida, Merlino's lawyer] were one hundred percent true," Borgesi wrote. "You know that and I know that. . . . There are a lot of things you wouldn't even believe. . . . It's fucked up how some things work out and especially with this stuff. I wish certain people would do the right thing, but I guess it's not in them. You were right about some things you said at 30th Street Station and like I said, you would really laugh at some things."

Borgesi went on to rip both Previte—"that rat fat fuck"—and Natale—"that lying, bald-headed, rat, drug-dealing faggot."

"That really is another story," Borgesi wrote of Natale. "You really wouldn't believe some of the stories."

Borgesi also wrote that his uncle, Joe Ligambi, said to say hello and to tell him that he occasionally saw Scafidi's mother around the neighborhood, and sometimes wrote to his brother, Tory, in prison.

"See, Tom, when I think of shit, it makes me sick. I'm sure everything will work out for everybody anyway. If you get me on your list, I'll come and see you. But I don't know if I can. . . . I'll send you some commissary. I gave Nella [Scafidi's wife] my beeper number so if she needs me or something, to call me. Forget that other shit. It's all bullshit."

In six years, Scafidi would tell a jury—the bitterness still evident—no one had made any effort to talk with or help his wife. As far as Borgesi, Merlino, and the other guys he had grown up with were concerned, "I was dead. . . . I wasn't even a person."

Now, with Merlino in jail, with Previte and Natale cooperating and the feds tightening a noose around the organization, now all of a sudden Borgesi claimed he wanted to mend fences and talk about the old days.

Scafidi said he wasn't buying. "Georgie loved Joey," Scafidi said. "I didn't believe anything he was telling me."

"Forget that other shit," Borgesi wrote. "I'll tell you the exact story about stuff. And the rat [Natale], he can't say anything about you, me, or anything else. We never did anything like they think, anyway."

Borgesi ended the letter with references to the happier times they'd spent growing up in South Philly.

"All you worry about is gin and Jimmy Roselli songs," he wrote to Scafidi. "Well, like I said, there are a lot of things I would like to tell you in person and I will. You could write back if you want, and I will give you the number of the macaroni place to call me. Take care, George.

"PS—Remember . . . [that girl] with nothing but a fur coat on?"

Less than a month after he got the letter from Borgesi, Gaetano "Tommy Horsehead" Scafidi signed a cooperating agreement with the U.S. Attorney's Office for the Eastern District of Pennsylvania. The deal was sealed on March 24, 2000.

Scafidi admitted to his role in the attempted murder of Joe Ciancaglini Jr. on March 2, 1993, and said he drove a backup car in a hit on James "Jimmy Brooms" D'Addorio, a neighborhood bigmouth killed on May 29, 1992, because he was harassing and disrespecting members of the organization.

Scafidi faced a potential sixty-year prison sentence, but his plea agreement carried a promise that the feds would speak on his behalf at the time of sentencing if he testified honestly and truthfully for the government.

Six days later, Mayor Milton Milan was indicted in Camden on corruption and racketeering charges. The nineteen-count federal indictment, announced by the U.S. attorney for New Jersey, accused Milan of selling his office to the mob. He was also charged with soliciting bribes from vendors who had contracts with the city, laundering money for local drug dealers, trading official favors for gifts, and skimming from political campaign committees.

Natale was clearly the source for the three counts in the indictment that outlined Milan's dealings with the Philadelphia branch of La Cosa Nostra. Among other things, the indictment charged that "Milton Milan accepted payments of money, travel expenses, and other benefits from associates of the Philadelphia LCN Family." The payments, the indictment alleged, were made at the direction of Natale. And "in exchange . . . on a continuing basis . . . Milan agreed to assist in obtaining contracts and other business with the City of Camden, and did official favors for persons associated with the Philadelphia LCN Family."

Danny Daidone, while not identified by name, was referred to repeatedly in the indictment as the mob's middleman in its dealings with the mayor, the conduit for payments "in excess of $30,000." Tapes made by Previte were also used to bolster the case.

In government documents that surfaced as Milan moved toward trial, the FBI outlined conversations in which Natale and others talked with Previte about what they hoped the mayor could do for the organization.

In one conversation shortly after Milan was elected mayor, Tony Viesti described the new mayor for Previte. Viesti, who was never shy when it came to grabbing a buck himself, complained that Camden's chief executive was constantly taking money from Natale. "He's got his hands in his pocket," Viesti said, then added, "[and] his arms and his legs."

"He's greedy?" Previte asked.

"Oh, yeah," Viesti replied.

In another conversation, Previte had pitched an idea to Natale that involved Camden. He had a friend who wanted to run concerts at the new Entertainment Center down on the Delaware riverfront across from Philadelphia. The idea was to stage the concerts for some charity, with the mob grabbing a piece of the action. Natale liked the idea and said his connections with the mayor could make it happen.

"The guy, the mayor, sweetheart this guy, you'll love him," Natale said of Milan. "Now I can't be seen with him anymore . . . but Danny Daidone, he loves Danny. He loves him. Fuhgeddaboudit. He loves him. He'll do anything in the world for me. I told him, he wanted to come over the night of the [inauguration party]. I said, 'You crazy. I can't be seen. . . .' I took care of him. I give him three thousand dollars in the envelope. I always took care of him. Give it to Danny. Gives it to him. Now they're gonna meet, I think next week. . . . I'm gonna give him five. . . . It assures us that he's our kind of guy. He's a 'man's man.' Smart . . . And looking to make money."

Milan, displaying an arrogance and bravado that would

hold up until a jury delivered its guilty verdict several
months later (by which point he was clearly shaken, and appeared to lose control of himself), posted a one-hundred-
and-fifty-thousand-dollar bond, was released from federal
custody, and promptly took the next day off. That Friday
morning he was playing golf at the Ramblewood Country
Club in nearby Mount Laurel.

"I'm probably hitting them the best I ever hit them," a
smiling Milan told an Associated Press reporter. "Especially
my short game. I'm just here to relax."

By then, at least in underworld circles, he was old news.

On March 31, 2000, while Milan was working on his golf
game, the U.S. Attorney's Office in Philadelphia unveiled a
second superseding indictment in the Merlino case. This
was the one everyone was expecting.

In addition to repeating the cocaine-trafficking charge
against Merlino and the stolen property charges added in
December, the new thirty-seven-count indictment included
multiple murder and attempted murder charges along with
counts of gambling and extortion. All were placed under the
umbrella of the RICO Act. The list included racketeering
and racketeering conspiracy charges and a litany of so-
called predicate acts—crimes ranging from murder and attempted murder to gambling, extortion, and the receipt of
stolen property—used to support the racketeering allega-
tions. There were also dozens of "substantive acts," specific
charges for individual crimes.

The murders included the Billy Veasey, Joe Sodano, and
Anthony Turra hits; the attempted murders included the Joey
Ciancaglini and Anthony Milicia shootings. There was also
a broad murder conspiracy charge that encompassed the war
with the Stanfa faction of the mob and the ambush on the
Schuylkill Expressway.

The entire hierarchy of the organization was named. Merlino, Borgesi, Mazzone, and John Ciancaglini all faced murder charges. Other defendants named were Ralph Abbruzzi,

Frank Gambino, Marty Angelina, Angelo Lutz, Anthony Accardo, Steven Frangipani, and Stephen Sharkey.

The extortion victims included Michael Casolaro as well as
several other young bookmakers from Delaware County
whose dealings with Borgesi and Lutz were picked up on the
state police wiretaps. Another extortion victim named in the
indictment was the owner of a video poker machine distribution company, who said that Ciancaglini and Angelina had
forced their way into his business. Angelina was also charged
with receiving a stolen Lamborghini in the deal set up by
Natale.

The Lamborghini deal epitomized the mob of Natale and
Merlino.

Natale had a mob friend from Cleveland whom he had
met in prison. The wiseguy got involved in a car theft ring,
a *Gone in 60 Seconds*–type operation.

The Lamborghini, a 1988 fuel-injected 500S model anniversary edition with gold wheels and only five thousand
miles on the odometer, would have been worth between
ninety-five and one hundred and ten thousand dollars if purchased through a dealership. The problem was, its theft was
such a high-profile job, the mobsters in Ohio couldn't move it.

In the end, Natale's friend offered to give it to him. Natale, in turn, brought Angelina and Merlino into the deal.

"I gave it to Marty Angelina 'cause he knew a lot about
stolen cars and Joey Merlino knew a lot about car dealers,"
Natale later explained. He said he told Angelina that he and
Joey could split whatever they got for the car.

"It's yours," he said.

The Lamborghini sat in a garage owned by Stevie Mazzone for nearly a year while Merlino and Angelina tried to
find a buyer. When they finally got a nibble, they asked for
thirty thousand dollars. The buyer offered twenty thousand.

They took the deal.

The buyer happened to be an undercover FBI agent from
Cleveland who had been tracking the car theft ring. As part

of the negotiations, Angelina promised to obtain a phony
title and license plates for the vehicle. He also said he would
replace its battery, which had died while the car was sitting
idle in Mazzone's garage.

More negotiations followed. Angelina never got the fake
title and tags. And at one point, spooked by the presence of
some Philadelphia cops near the garage, he and an associate
tried to move the vehicle. Instead of a flatbed truck, however, they rented a Ryder panel truck. They tried to get the
Lamborghini into the vehicle by driving it up a makeshift
ramp made of wooden planks. In the process the car—a limited edition, six-figure piece of machinery—slipped off the
ramp and crashed to the ground.

"They dropped the car," the undercover agent later said
incredulously.

With a now-damaged bumper and still without any tags
and title, the Lamborghini was eventually delivered to the
undercover FBI agent. By that point, the price had plum-
meted to a laughable fifty-three hundred dollars.

Angelina took the cash. The feds took the car. Now the
entire episode was part of a racketeering indictment.

"Not only does it reach into the highest levels of organized crime," Michael Stiles, the U.S. attorney for the Eastern
District of Pennsylvania, said in announcing the new indictment, but "significant evidence comes from the highest levels of organized crime. That has never occurred before."

Stiles was clearly referring to Natale, whose cooperation
and information provided the basis for the charges of murder and attempted murder that had been added to the case.

Natale's decision to cooperate, Stiles said that day, "rep-
resents a complete collapse of this criminal organization."

And yet there was more to come.

Sporting a new goatee and clean-shaven head, Ralph Natale
confidently strode into a federal courtroom in Camden on
May 5, 2000, to formalize his deal with the government.

During a ninety-minute hearing, he answered a series of questions posed by Assistant U.S. Attorneys Mary Futcher of New Jersey and Barry Gross of Philadelphia, pleading guilty to two federal informations—charges brought in lieu of an indictment.

In the New Jersey case he admitted his role in the methamphetamine distribution ring, in attempts to corrupt and bribe officials in Camden, and in the murder of Joe Sodano. In the Philadelphia case, he pleaded guilty to murder and attempted murder charges and assorted gambling and extortion offenses. All were part of his life in La Cosa Nostra, he said.

Responding "Yes," "Yes, sir," and "I certainly did," Natale was the embodiment of cooperation, offering a total and complete confession to crimes that stretched from 1970—when he said he killed a union rival—through 1998. In all, he admitted to his own involvement in eight murders and four attempted murders. He also detailed how he had bribed Milan and schemed to grab city contracts for the mob.

Dressed in what could have been a tailor-made blue business suit, crisp white shirt, and blue tie, Natale clearly enjoyed his new role and his time in the spotlight. He waved to his wife, one son, and dozens of federal authorities—FBI agents and prosecutors—who had crowded into the courtroom for the historic denouement.

The only two people in the courtroom who seemed less than impressed were John Ciancaglini's wife, Kathy, who had shown up out of curiosity, and Mike Pinsky, at the time the defense attorney for Borgesi. Kathy Ciancaglini still couldn't believe Natale was doing what he was doing. Her husband, she contended, had brought legitimate business deals to Natale, who was always bragging about the projects they were going to launch. None of it ever happened, she said. And now this.

Pinsky, part of a team of defense lawyers who were preparing for the racketeering case, sat in the second row of the fifth-floor courtroom taking notes. As he left court that day, he was asked why he had come to the hearing. "I was here watching the government make a deal with a serial killer," the veteran defense attorney quipped.

Natale's lawyer, Marc Neff, told reporters after the proceeding that his client was "comfortable" with his decision to cooperate and was ready to testify.

Later that same month, both Pete the Crumb Caprio in Newark and Bobby Luisi in Boston cut deals with the government. Both agreed to cooperate in exchange for leniency at the time of sentencing. Luisi, of course, had been indicted for cocaine trafficking. Caprio, in an indictment handed up in Newark early in March, had been charged with two murders, including the Sodano hit. The charges were based on information being provided by Philly Fay Casale and Ralph Natale.

As the summer of 2000 approached, with Merlino and most of his top associates in jail (all the principal defendants had been denied bail), federal authorities had a list of mob informants ready to take the stand. These included Natale, Previte, Scafidi, Luisi, Casale, and Caprio.

But unlike the other mob racketeering and murder cases that had rocked Philadelphia in the past—the Scarfo trial in 1988, which included nine murders and four attempted murders, and the Stanfa trial in 1995, which included five murders and nearly a dozen attempts—the only murder in the current indictment for which the prosecution had an eyewitness account was the Joe Sodano hit.

The shooter, Philly Fay Casale, was prepared to testify that he blew Sodano away on Caprio's orders. Caprio, in turn, was prepared to testify that he had set the hit in motion after getting the okay from Natale and Merlino. Natale was prepared to testify that he and Merlino approved the hit because of Sodano's continued refusal to come to meetings when called.

According to the government's theory of the case, four

people were involved in the plot to kill Joe Sodano. The government had made deals with three, including the shooter, in order to charge Merlino with the slaying.

Defense attorneys rolled their eyes.

Something was wrong with that math.

18

Ralph Natale's debut as a prosecution witness came in the fall of 2000, when he took the witness stand in the corruption trial of Mayor Milton Milan. Although he only testified for parts of three days in a trial that lasted more than five weeks, his appearance was one of the high points of the prosecution.

This was, after all, a sitting mob boss taking the stand, renouncing La Cosa Nostra to offer the government unprecedented access to the inner workings of the most secret society.

Natale played the part to the hilt.

Again a fashion plate in well-cut business suits, bright white shirts, and color-coordinated ties, he was by turns bold and humble, brash and self-deprecating. When he told the story of how and when he decided to cooperate—the "epiphany" that occurred when he saw his long-suffering wife and family sitting in the courtroom at his arraignment in the drug case—his voice was quiet and somber, dripping with pathos.

But when he talked about his role in the mob, from the Bruno years through the Merlino era, he was full of bravado.

He was Bruno's "dog," he said, ready to attack on command. After he was convicted in the 1970s for drug trafficking and arson, he did his time, keeping his mouth shut, honoring the code of *omerta* even though he hadn't been formally initiated into the mob.

Later, Natale would claim he had turned down Bruno's offer to become a member of the crime family because he knew he wouldn't get along with those who surrounded the mob boss. Sooner or later, he said, either he'd get killed or he would have to start killing them. He was singularly attached to Bruno, a man he trusted and respected. But he felt nothing but disdain for several of Bruno's lieutenants.

That everyone from that era was either dead or in jail and in no position to refute what he was saying was typical of most of Natale's mob talk. But the jury appeared mesmerized by his testimony. It was, in many ways, their own episode of *The Sopranos*.

Natale said he became boss at the urging of Joey Merlino. "At the time, there was a gang war going on between John Stanfa and a group of young men from South Philadelphia," he said. "They came to me for support . . . and at that time I was being incarcerated with Joey Merlino and he asked me to help him because of my prestige and what he knew about me."

He said he was the "unofficial" boss of the Philadelphia mob while in prison, and that after he was released in 1994 the New York families happily and quickly recognized him. "A group of young men that I supported in prison rallied around me," he said. "We took a trip to New York. It was confirmed by the other families in New York. There's five families in New York City. They said they would only recognize one man in Philadelphia after the fiasco that happened after Nicky Scarfo.

"They didn't recognize Philadelphia until I decided to become the boss. . . ."

This was all delivered in a quiet and emphatically sincere voice. He went on to explain that he was able to get the crime family up and running by borrowing cash from businessmen and others who were anxious to be associated with him.

They readily parted with their cash, he said, "because of . . . my prestige as a gangster and a man.

"I went out and borrowed money from people that I knew who were basically legitimate," he said. "I also tried to influence quite a few other people on lending us money, which they did, and we put together a little bankroll and we started the business of crime again in the city of Philadelphia and South Jersey. . . . Nobody was freelancing no more. Everything illegal in that city belonged to us, La Cosa Nostra."

As he left the courthouse following Natale's first day on the witness stand, Milan told reporters, "The devil himself came to testify."

Milan, of course, had more to worry about than Natale. Before it was over, politicos, drug traffickers, legitimate contractors, and city workers would all take the stand helping the government build its case.

But Natale set the tone. He was the star witness. And it was clearly with that in mind that the government called him early in the trial.

That the jury was going along with the show was evident on the second day, when Natale's quips and asides brought smiles and laughter to those sitting in the box.

When he had trouble recalling some dates, he apologized. "If I thought I was going to be up here, I would have marked it down," he said.

He also sheepishly admitted that the precautions he had taken to avoid wiretaps had failed miserably. He was, he said, constantly on guard for government surveillance. And, he said with a self-deprecating grin, his concern was justified. "I was right," Natale said, after being questioned from the transcripts of several tapes that had been played for the jury. "We're reading it right now."

Natale appeared to enjoy the attention, the jurors hanging on his every word, the prosecution leading him with questions to which he already had the answers, the packed courtroom buzzing with his every clever remark.

He described Milan as "the golden goose" who was going to lay "the golden eggs" for the mob. "He had a girlfriend, he had a life," Natale said of the mayor, sounding like a wise older man of the world who understood such things. "He needed this. He needed that. . . . Whatever it was, I said I'd put it in an envelope. . . ."

Natale detailed how he had funneled thousands of dollars in cash and gifts—most of it through Daidone—to the mayor. He knew Daidone, he said, from his days with the Bartenders Union. Daidone's brother, Albert, had been an organizer for Natale.

Danny Daidone, he said, "was my go-between" in Camden. "Danny had that appearance as a businessman. He could speak much better than a lot of the hoodlums that I had around me. I said he would be perfect for me. And that's what we did."

Natale also testified that any time he sent money to Milan, it was in the form of hundred-dollar bills. "I wouldn't insult the man by giving him twenties," he said. And of course the payments had to be in cash: "I can't write a check out from La Cosa Nostra," he said, drawing laughter from both the jury box and the courtroom.

Two women who watched the "performance" that day, however, appeared less than impressed.

Kathy Ciancaglini, John's wife, said the deeply tanned Natale, with his salt-and-pepper goatee and cleanly shaved pate, looked like "Daddy Warbucks." His story, she said, was "unbelievable." Shaking her head and rolling her eyes, the attractive but volatile Mrs. Ciancaglini made it clear that she meant that in both the figurative and literal sense.

Sitting in the same aisle that day in court was a neatly dressed blonde who watched Natale's every move. He made no indication that he noticed her, but it's hard to imagine he did not.

Ruthann Seccio declined to comment on the testimony of her former boyfriend. Asked if she had anything to say, she replied with a terse "No," then made a motion as if sealing her lips shut.

Edwin Jacobs Jr., who had recently been retained to represent Merlino in the racketeering case, was also in court for Natale's testimony. The high-profile New Jersey criminal defense lawyer wanted to see Natale in action. Jacobs had once represented Merlino's father, still represented his cousins, and had successfully defended lawyer Sal Avena in the Stanfa case (it was in Avena's law offices that the FBI had taped hundreds of mob conversations). Jacobs was a master at studying and using tapes to his advantage. Like the Stanfa case, the pending racketeering trial was loaded with secretly recorded conversations.

As expected, Jacobs came away less than impressed by Natale. "This is a man who, by anybody's yardstick, is a monster," he said. "He's a self-admitted murderer, drug dealer, and extortionist. But they've dressed him up, rehearsed him, and they've got him on a tight leash. They're presenting him as a civilized, white-collar criminal.

"It's good theater. But it's fiction."

Whatever it was, the jury seemed to both enjoy and believe it.

The government, of course, used other evidence to support Natale's claims that he had bought the mayor of Camden. There were, for example, the travel agency records for the mayor's trip to Florida, paid for by Daidone. There were surveillance videos and FBI agent testimony of meetings between Daidone and Milan. And there were tapes, lots of them: tapes of Natale boasting about his plans to win government contracts, tapes of Natale, Daidone, and others bitching and moaning when the contracts never came through.

In one classic conversation, Daidone complained to a business associate about all the things he had done for the mayor, including giving away turkeys to poor families in the city at Christmastime. Joey Merlino, it turned out, wasn't the only wiseguy with a soft spot for the needy, although after listening to Daidone on tape it's hard to consider him philanthropic.

This tape was played for the jury by Carlos Martir, Milan's defense attorney. Martir was trying to show that, despite Natale's claims to have the mayor in his pocket, the mob never received a contract from the city. "I haven't been given one contract in that city, not one," Daidone said in a phone call taped by the feds. "I'm fuckin' livid. . . . I'm handing out fuckin' money like it's going out of fuckin' style for every fuckin' thing they asked me to do in that city. Every fuckin' thing . . .

"I didn't have to give him no campaign money. I didn't have to buy fuckin' turkeys. . . . I didn't have to go out in the cold fuckin' weather like I did . . . and hand out fuckin' turkeys to fuckin' people with snot hanging out of their fuckin' nose. But I fuckin' did it. . . . And then be treated in this fuckin' way?"

The Milan trial took five weeks. The government called sixty witnesses and presented the jury with over seven hundred exhibits. The panel of eight women and four men deliberated for seven days before reaching a verdict.

Milan, who had celebrated his thirty-eighth birthday during the trial, was convicted on fourteen of nineteen counts. He was acquitted of two counts of attempting to extort a campaign contribution, and the jury hung on three counts of mail and wire fraud.

The verdict was delivered a little after 3 P.M. on December 21, 2000. Milan's bail was immediately revoked. He was handcuffed and taken off to prison.

Outside the courthouse, Robert Cleary, the U.S. attorney for New Jersey, described Milan as a "one-man crime spree, someone who sold his office, someone who sold out the very people he was elected to serve."

Among other things, Cleary said, the trial had established the "credibility" of mob boss Ralph Natale.

One year after the February 1995 murder of Ralph Mazzuca, Roger Vella was preening like a South Philadelphia street-corner rooster when a reporter asked him his thoughts on the ongoing investigation.

Vella, who loudly and proudly proclaimed himself a friend and confidante of Skinny Joey Merlino, told the reporter he had a message for the police detectives who were working the homicide case. "What's taking so long?" he said. "If they're gonna arrest me, arrest me. Bring it on. I ain't afraid. Bring it on!"

It took four more years, but the cops did just that.

Vella was arrested for the Mazzuca murder on October 21, 2000. The police came to his home early on that Saturday morning. He's been in jail ever since. On December 13, just as the jury was beginning deliberations in the Milan case, Vella was arraigned in Common Pleas Court in Philadelphia on first-degree murder charges.

Investigators provided only a bare-bones outline of the case they had against the then twenty-nine-year-old mob sycophant. Vella, they alleged, shot Mazzuca because he suspected him of being behind the home invasion in which his family was tied up and pistol-whipped. Then, with the help of "others," he allegedly dumped Mazzuca's body near the Food Distribution Center before setting it on fire.

One witness called to testify at the hearing, a former friend of Vella's, said that Roger had bragged about the murder, claiming "Joey" had approved it. Another witness, who was working at a gas station that night, recalled selling a dollar's worth of gasoline to someone resembling Vella shortly before the body was discovered in flames.

Investigators, of course, had more, but that was enough for the judge to order Vella held without bail pending trial.

Both Natale and Scafidi provided information about the hit. Both clearly linked Vella to the murder. But their information conflicted on the issue of whether the shooting was mob-related.

Scafidi told the FBI that he had been in prison with Vella briefly in the late 1990s, when Vella was doing time for a drug conviction. According to an FBI report, Scafidi said that Vella had bragged about the hit and "said he was going to be proposed for membership in the Philadelphia Family of La Cosa Nostra by George Borgesi" because he had murdered Mazzuca.

Natale also told the FBI that Vella had killed Mazzuca. He said that Borgesi told him about the murder after the fact, and had said that he helped Vella dispose of the body. According to that FBI report, however, Natale said the Mazzuca murder "was not a sanctioned hit."

If Natale was telling the truth, then investigators were heading in the wrong direction with their attempts to turn the Mazzuca homicide into a gangland murder and to put Merlino and Borgesi behind it.

Within months Vella would give his own version of the events, adding several more twists and turns to the story—a gritty South Philadelphia street-corner drama built around money, drugs, and revenge.

On February 27, 2001, Louis Barbone, the defense attorney for Jimmy DeLaurentis, stood before a federal jury in the same courthouse where Natale had testified and where Milan had been convicted, and blasted away at Ron Previte's credibility.

"Ron Previte went about his life deceiving, stealing, and making every opportunity turn to his benefit," Barbone said in his opening argument in the bribery and extortion case of the Hammonton police detective.

The description, Previte later admitted, was dead-on accurate.

"This is a story of deception, thievery, plots, and schemes," Barbone went on. "It's about money. . . . Opportunity is Mr. Previte's forte."

The DeLaurentis case, which opened two months after Milan's had ended, was another chance for the government to take a key witness for a "test drive" and another opportunity for the defense to size up its opposition. While the trial did not attract the same kinds of crowds or media attention as the Milan prosecution, it did offer another preview of the bigger racketeering case—*United States v. Merlino*—that was scheduled to begin in U.S. District Court in Philadelphia within a month.

Barbone was the partner of Eddie Jacobs, who had taken over Merlino's defense. Barbone's representation of DeLaurentis gave the law firm the chance to be actively involved in the cross-examination of Previte rather than merely observing the process. Previte, they knew, would be a key to the racketeering trial.

Barbone had already opened another front in the attempt to undermine Previte. In the fall of 1999 he filed a claim on behalf of Michael Perna, a former Hammonton insurance broker, who claimed that Previte had extorted thirty thousand dollars from him and caused him to leave town, in fear for his life, moving to Las Vegas.

The claim, which was later expanded to a civil action, sought over one million dollars from the FBI and New Jersey State Police, alleging that the agencies had allowed Previte to run wild while working as a law enforcement informant. It was a theme that would come up again and again when Jacobs cross-examined Previte in the Merlino trial.

Perna said he was the target of a shakedown in 1993 and 1994 because his insurance company did business with the casino industry. He said he was told he would have to pay the mob in order to keep writing policies in Atlantic City.

Previte, he said, threatened him and his family. After making two extortion payments—twenty thousand dollars on one occasion, ten thouand on another—Perna said he closed his New Jersey business and headed for Vegas.

No one in law enforcement, he said, would help.

In fact, he later alleged, a New Jersey State Police detective said there was nothing anyone could do because of Previte's ties to the FBI. "You're stuck in the middle here," the New Jersey investigator allegedly told him.

In the lawsuit, Barbone claimed that the law enforcement agencies knowingly allowed Previte to commit crimes while on the government payroll. He alleged that federal authorities "did . . . conceal, protect, and otherwise provide Previte with a badge of authority upon which to perpetrate crimes against [Perna] . . . knowing of Previte's illegal and illicit pursuit for his own economic benefit." The suit also claimed that the New Jersey State Police had failed to protect Perna after he told them of the extortion.

While it was eventually dismissed by a federal judge, the suit provided the defense with another chance to take shots at a key government witness.

Previte, like Natale, seemed to enjoy his time on the witness stand. As a former police officer, he was no stranger to testifying. He spent three days being questioned by the prosecution, recounting his life of crime, his years of cooperating with authorities, and his decision to wear a wire for the FBI.

Among other things, Previte had to provide details about his financial arrangement with the feds. When he agreed to strap on a wire, he said, he signed a "service contract" with the FBI. The contract, which was shown to the jury, provided that Previte would be paid $8,610 a month. Signed in February 1997, the arrangement was still in effect, monthly stipend and all, when he took the stand in the DeLaurentis case.

Previte also acknowledged that as part of his deal with the government, he pleaded guilty to extortion in the Hammonton case, admitting that he had shaken down a bar owner whose liquor license was in jeopardy. He faced about thirty months in prison as a maximum sentence, but conceded that the government had agreed to speak in his defense at the time of sentencing. He expected to be given probation, no jail time.

Previte was not charged with any other criminal offense.

He got a pass on everything else—the bookmaking, the shakedowns, the prostitution, and the drug dealing. His plea wiped the slate clean.

Under the terms of his service contract, the FBI also picked up the tab for Previte's car, his car insurance, his life insurance and health care. And he was in line for a one-time bonus payment once his work as a cooperating witness was officially terminated.

In all, including the lesser payments he had received from both the New Jersey State Police and the FBI in the years leading up to his time on the wire, Previte had been paid about five hundred thousand dollars. The bonus, which came a couple of years later, pushed the figure to the one-million-dollar mark.

Previte also acknowledged that he paid no income tax on any of his earnings while working undercover. That was another area that the defense would hammer away at whenever he got in front of a jury.

It was all standard defense rhetoric—the kind of arguments that crop up in any mob case in which a cooperating witness has received any kind of deal. The defense hopes that the jury—a panel of twelve men and women who don't routinely come in contact with the criminal justice system—will be surprised and outraged. In most cases, jurors are not.

Still, the game gets replayed with each new trial.

Both Natale and Previte were getting out from under serious criminal charges by cooperating, the defense contended. That alone was reason enough to question their testimony: surely they would say whatever the government wanted to hear. Previte was also being paid by the very people who were presenting him as a credible and believable witness. If we tried to do that, the defense contended—if we offered witnesses money—no one would believe anything they said, and we would be arrested for obstructing justice.

Previte either ignored the taunts or threw his standard verbal counterpunch. Each day he put on that body wire, he said, could have been his last. This was a cold and ruthless crime family, a mob that time and again had murdered its own.

Previte had seen it all firsthand during the Stanfa-Merlino war. "Everybody was fighting over power and money," he said. "People were getting shot. People were getting stabbed. People were dropping like flies."

Though he'd experienced all that up close, Previte said, he nevertheless agreed to put on a wire for the FBI. Any money he was paid, he earned, he said emphatically. It was "hazardous duty pay."

While Previte at times sounded rehearsed while answering questions posed by Assistant U.S. Attorney Mary Futcher, he seemed to come alive under the cross-examination of Barbone.

When the defense attorney began questioning him about his career as a bookmaker, Previte readily admitted that he'd been taking bets for years in the Hammonton area. When Barbone asked how much action he would take, Previte said he had developed a "feel" for his customers. He said it was important to keep gamblers from exceeding their limits. He said he took bets that ranged from ten dollars to twenty-five hundred. On occasion, he said, he would take a bet for as much as ten thousand dollars. It would depend on the customer.

Then he paused, looked at the jury, and looked back at Barbone. "Someone like you," Previte told the defense attorney, "could probably afford to put a ten-thousand-dollar bet on."

The jurors, the judge, even Barbone, laughed.

"Touché," the lawyer said.

"Anything that was entrusted to your care, you'd steal," Barbone said at another point, more a comment than a question.

"Yes," Previte replied unashamedly.

Barbone had tried throughout the trial to argue that DeLaurentis was set up by Previte, that the detective was friendly with the mobster but was not his partner in crime.

Prosecutor Mary Futcher, on the other hand, echoed the position laid out in the Milan case. DeLaurentis, the detective who was in charge of liquor licenses for the town of Hammonton, was "a corrupt police officer, an arrogant police officer who sold his job, who extorted those people he was supposed to be protecting."

In all, the six counts against DeLaurentis charged that he had used Previte to extort fourteen thousand dollars. The jury found the detective guilty on two counts, one each of bribery and extortion. The crimes involved the six thousand dollars Previte had collected *after* he began wearing a wire for the FBI—the bribery and extorting that he and DeLaurentis openly discussed on tape.

DeLaurentis was acquitted of three counts of extortion and one count of bribery, involving eight thousand dollars he allegedly collected with Previte before Previte began taping for the feds. Previte's credibility, some defense attorneys decided, would only be as good as his tapes. For Merlino and those awaiting trial, that was a mixed message.

The DeLaurentis verdict was handed down on April 10, 2001. By that point, Ralph Natale was already on the witness stand in federal court in Philadelphia.

19

If you were looking for an omen, it was right there on the calendar.

The first jurors in the Joey Merlino racketeering trial were selected on March 21, 2001, the twenty-first anniversary of the murder of Angelo Bruno. On the night Bruno got it, Joey Merlino was eighteen. Back then, the Philadelphia mob was an established, quietly run, and highly efficient organization.

This trial would demonstrate what it had become.

Bruno, who believed in making money, not headlines, had set the tone during a period of Mafia wealth and prosperity that stretched from 1959 to 1980. Most of those who worked under him fell in line. Then a guy in a raincoat walked out of the shadows, pulled out a shotgun, and blasted a hole in the back of Ange's head.

Twenty-one years later, it didn't matter if the shooter that night was Tony Bananas Caponigro or someone he had sent. It was of little consequence whether the hit was set in motion in a dispute over drugs, Atlantic City, or control of the crime family. The only thing that was certain was that the shotgun blast that punctured that quiet South Philadelphia night marked the end of Angelo Bruno, and the beginning of the end of his organization.

Bruno had been killed as he sat smoking a cigarette in a car parked in front of his modest row house at 934 Snyder Avenue in the heart of South Philadelphia. The trial that was about to play out in the ninth-floor courtroom of U.S. District Court Judge Herbert Hutton completed the dismantling of his mob.

It would take five months, in a relatively long and sometimes rambling proceeding that ultimately provided a startlingly accurate snapshot of La Cosa Nostra circa 2000.

Where to begin?

Perhaps on the day, early in the trial, when there were three different mob bosses in court, one on the witness stand, one at the defense table, and one in the audience. There was Ralph Natale, full of bluster and bravado, testifying for the government. There was Joey Merlino, his dark eyes flashing as he sat at the defense table whispering in his lawyer's ear, trading notes with codefendants George Borgesi and Steven Mazzone, or asking Angelo Lutz for a piece of candy from the stash the defense shared throughout the proceeding. And there was Joe Ligambi, described by the feds as the acting boss, dressed in a dark black leather jacket, sitting with the group of the defendants' wives, girlfriends, sisters, and brothers who came to court each day.

No one could ever remember a mob trial—in Philadelphia or anywhere else in the country, for that matter—where three bosses sat in such decidedly different positions. Sure, there had been trials in which mob bosses had been defendants together; the Mafia Commission trial, prosecuted in New York in 1986 by Rudy Giuliani and the U.S. Attorney's Office, was perhaps the most famous. And there had been other trials in which former or acting bosses had testified for the government. Cleveland's Angelo Lonardo had been a key government witness in the 1970s and early 1980s; Alphonse "Little Al" D'Arco, the acting boss of New York's Lucchese family, began cooperating in the 1990s.

But only in Philadelphia—only with this mob of misfits, malcontents, and mopes—would three Mafia dons from the *same* family be in court at the same time for different reasons.

Merlino and Natale, of course, had to be there. But Ligambi, the sixty-two-year-old gangster who had done ten years on a murder conviction that was later overturned, usually shunned the limelight. Maybe he felt he should be there to support his nephew, Borgesi, who was one of the defendants; or his sister, Borgesi's mother, a frequent court spectator. Maybe he just couldn't stay away.

At first Ligambi was surprisingly open and talkative. "I knew him from when he was a bartender at the Rickshaw [an old Cherry Hill restaurant]," Ligambi said as he stood on line the day Natale first took the stand. "He's a couple of years older than me. He was nothing. . . . Then when he was telling these kids all these stories I used to say, 'What? It's all bullshit.'

"It's all lies."

Ligambi didn't like seeing his comments in the newspaper the next day, and quietly asked that anything else he said be off the record. Once he realized that his daily appearances afforded the media a chance to take his picture and allowed the feds to watch him interact with other mobsters and mob associates, he stopped coming.

But in the beginning, Ligambi, who took over the crime family after Merlino was arrested, would line up with dozens of other spectators in the hallway outside Hutton's courtroom each morning. He was usually accompanied by Joseph "Mousie" Massimino, a short and hyperkinetic bookmaker and convicted drug dealer who, according to the feds, soon emerged as the new underboss of the bedraggled Philadelphia mob.

An underboss named "Mousie"—that was all anyone needed to know about the new Philadelphia crime family.

Jury selection took a week. Twelve jurors and six alternates were chosen to hear testimony. All were selected anonymously. By that point, four of the original eleven defendants—"Ralphie Head" Abbruzzi, Steven Frangipanni, Anthony Accardo, and Stephen Sharkey—had pleaded guilty and were out of the case.

Those four were looking at sentences of from two to five years, depending on the nature of their plea. All would get a break, under sentencing guidelines, for admitting their guilt. None was cooperating.

Abbruzzi and Frangipanni had pleaded guilty to their roles in the stolen property ring in August 2000 and were awaiting sentencing at the time the Merlino trial started. Accardo and Sharkey pleaded guilty just days before the jury selection process began.

Accardo, who was often referred to as "Tony Cugino" (Cousin Tony), admitted his role in the stolen property ring, and also confessed to taking part in a brutal extortion in October 1999 in which he terrorized a deadbeat gambler who owed the mob fifty-five hundred dollars. Accardo, according to tapes the government was prepared to play and testimony from witnesses who would be called to the stand, used a billy club to beat the gambler; then, in a follow-up phone call, he told his hapless victim that if he didn't come up with the cash by the following Friday, "the undertaker" would take him out.

The beating that Accardo administered, federal prosecutors said in a plea memorandum, was so severe that the victim "defecated in his pants."

Sharkey, a close friend of Borgesi's, pleaded guilty to gambling and extortion charges.

By the time the lawyers offered their opening arguments, Angelo Lutz was the only mob associate still in the case. The other six defendants—Merlino, Borgesi, Mazzone, Ciancaglini, Angelina, and Gambino—were all made members of the mob, according to the prosecution.

Lutz had balked at a deal that would have required him to admit his role in an extortion plot, a deal similar to Sharkey's. "I'm a gambler," he said repeatedly in the months leading up to the trial. "I'm not pleading guilty to something I didn't do."

It would be Fat Ange's biggest gamble. Had he pleaded

guilty, he would have faced a sentencing guideline range from about five to seven years. Instead he chose to roll the dice, to take his chances with the jury.

The trial started at nine-thirty each morning. Spectators would begin forming a line in front of a metal detector just outside the courtroom doors about an hour earlier. The media were guaranteed the twelve to fifteen spots in the first row of the high-ceilinged courtroom. Spectators, on a first-come first-served basis, got the other four rows. But it quickly became the custom that family members, particularly wives and mothers, got to move to the front of the line no matter when they arrived. In all, about seventy people could fit in the courtroom. On most days there was a capacity crowd, with a dozen or more people waiting outside for a chance to grab a spot if anyone left early.

People seldom did.

Merlino, who had lived his entire life in the spotlight, was going to go out the same way. He had created a buzz wherever he went: the hip clubs on Delaware Avenue that catered to upwardly mobile thirtysomethings, the bars at the shore that were packed with both yuppies and blue-collar vacationers each weekend, the ballparks and stadiums where the Phillies, Flyers, and 76ers played. Now that same buzz, that same sense of anticipation, had been transplanted to the ninth-floor courtroom.

Joey's in the house.

The prosecution team was made up of four lawyers, each of whom brought something different to the table. The lead prosecutor was Assistant U.S. Attorney Barry Gross, who had been trying mob cases since the Scarfo era. Zane Memeger, a young attorney recently added to the Organized Crime Bureau, was the number two man. Considered an up-and-comer in the Eastern District, he would offer the government's opening arguments. This was his first big case. Several people in the defense camp figured that the fact that

Memeger and the judge were both African-American gave the prosecution another advantage.

Assistant U.S. Attorney David Fritchey, another veteran of Philadelphia mob prosecutions, sat in the third chair; Steve D'Aguanno, on loan from the Philadelphia District Attorney's Office, was the fourth member of the team. Young and aggressive, D'Aguanno came from South Philadelphia, and knew his way around the neighborhood as well as anyone at the defense table.

For the defense, Edwin Jacobs Jr., the high-priced Atlantic City attorney representing Merlino, was the lead lawyer. Jacobs, always thoroughly prepared, had spent months going over the tapes and transcripts, the FBI reports, and the grand jury testimony. He knew the government's case better than some of the prosecutors. The knock was that he often went to extremes to demonstrate that. His style—he never made a point once that he didn't make three times— eventually wore thin with some of the other defense attorneys, especially those with as much criminal trial background as he had.

Sitting uncomfortably in the second defense chair was Bruce Cutler, the New York attorney who had defended and befriended John Gotti. Gotti's losing battle with cancer would become public during the trial, and may have been a distraction for Cutler, who seemed to be off his game. Cutler seemed ill at ease in the courtroom, and in the city of Philadelphia. He was there to represent Borgesi; so was Louis Natali, a local criminal lawyer and law school professor who served the defense team as an expert on appealable issues.

Mazzone was represented by Stephen Patrizio, another veteran criminal defense attorney. Ciancaglini's lawyer was F. Emmett Fitzpatrick, one of the deans of the Philadelphia criminal bar and a former district attorney. Angelina had Jack McMahon, an energetic and opinionated former assistant DA. McMahon, a master of the make-your-point-and-sit-down style, was uncomfortable with the minutely detailed defense launched by Jacobs. He thought this was an eight-week case. It would take fifteen.

Two court-appointed lawyers—Christopher Warren, who represented Angelo Lutz, and NiaLena Caravasos, who represented Frank Gambino—rounded out the defense team. Warren, considered a genius with a legal brief, quickly emerged as the one lawyer the others sought out for advice and strategy discussions. He would later be featured in an article in *Details* magazine about the case. Caravasos added a daily dose of calm, no-nonsense, just-the-facts lawyering to a legal proceeding otherwise overloaded with testosterone.

"La Cosa Nostra," said Memeger, struggling repeatedly with the Italian pronunciation during his opening statement, was set up to "instill fear, exert power, and make money." The young prosecutor would repeat that phrase a staggering nineteen times during his one-hour speech to the jury as he outlined the case against the defendants. He described the murders and attempted murders, the extortions, the shake-downs, the bookmaking, and the stolen property.

He acknowledged that the government would provide co-operating mob witnesses who were "not angels." But in a variation on the classic prosecutorial reminder that "swans don't swim in sewers," he said, "these defendants didn't deal with law-abiding citizens. They dealt with people like themselves."

Jacobs, presenting the first of the defense openings, sounded the theme that he and the other lawyers would come back to again and again during the trial. "We know what we did and we know what we didn't do," he said, tacitly conceding some of the gambling and stolen property charges that were captured so effectively on tape. But there was nothing other than the uncorroborated testimony of people like Natale and Caprio, he added, to support the murder and attempted murder charges.

"They all want a get-out-of-jail-free card," Jacobs said of the cooperators. "And there's only one in Philadelphia right now. That's Joey Merlino.

"You know who we are?" Jacobs asked, waving his arm in a sweeping motion that took in the entire L-shaped defense table. "We're the usual suspects."

And so it began.

Natale testified for fourteen days, stretched out over four weeks. He was questioned by the prosecution for six days, by Jacobs for six more, and by the other defense attorneys for the better part of two.

He repeated and expanded on many of the things he had said during the Milan trial, again recounting his epiphany, adding that as a government witness he had come over from "the dark side."

"I believed in every part of me in La Cosa Nostra," he said. "I didn't know the difference between right and wrong."

At another point he told the jury, "I've done everything wrong a man could do in life. I've broken every law of the government and every commandment the dear Lord ever made."

And when he retold the story about seeing his wife and family in court at his arraignment in the drug case—the moment when he decided "no more La Cosa Nostra"—he added a new bit of flare for the packed courtroom. Asked how he felt that day, he paused, and then said quietly, "I could have spit on myself."

Many of those who crowded into the courtroom, and most of those at the defense table, rolled their eyes. Most would have liked to spit on him as he sat in the witness stand. Cutler, in what may have been the legal high point of his stay in Philadelphia, had described Natale in his opening remarks as "a reprobate . . . a bum, a blowhard, and a fake." The defense spent most of its cross-examination trying to paint him just that way, trying to show that the information he brought to the table was without foundation.

Several attorneys noted that three of the men Natale had linked to mob hits, Gaeton Lucibello, Michael Lancelotti, and Michael Virgilio—were not defendants and had not been charged. How good could Natale's information be, the defense repeatedly argued, if men he alleged were killers were still at large?

Not only at large, but in that very courtroom.

McMahon desperately wanted to make that point, and on the day he cross-examined Natale he got into it again. Natale had placed Lucibello in the van as a shooter on the day John Stanfa was ambushed on the Schuylkill Expressway. He named Lancelotti as one of the gunmen who rushed into Joey Chang's luncheonette and opened fire. And he fingered Virgilio as the man who had shot and killed Anthony Turra.

Both Virgilio and Lancelotti were sitting in the back row of the courtroom when McMahon began to question Natale. Over the strenuous objection of the prosecution, Judge Hutton let McMahon point them out to the jury. Both were asked to stand. Virgilio, who according to earlier testimony suffered from Tourette's syndrome, shook uncontrollably as he faced the jury.

This, McMahon asked without uttering another word, was a mob hit man?

Natale, realizing what was going on, told the jury Virgilio usually didn't shake like that. He said he had probably "stopped taking his medicine."

Later, during a break in the trial, both men denied they had anything to do with the shootings.

"Absolutely not," said Lancelotti, when asked if he were part of the hit team in the Joey Chang shooting.

"No way," said Virgilio, still shaking, when asked about the Turra murder.

Then, moving his finger in a circle next to his head, Mikey Penknife said of Natale, "He's nuts."

Natale stuck to his story through the grueling cross-examination, shaking his head and smiling as he sought to maintain his composure while everything from his manhood to his memory was challenged.

Once, while discussing a loan-sharking operation and ticking off a series of figures, he held up his middle finger signaling "one," but also, to many who were in the courtroom, sending a "fuck you" message to the defendants.

The presence of Lucibello, Lancelotti, Virgilio, and a half-dozen other mob members and associates in court on a regular basis was another example of how the mob had changed over the years. During the Bruno era in Philadelphia and the Gambino era in New York, mobsters who weren't on trial would not be seen anywhere near an ongoing racketeering case. Why call attention to yourself, or to the proceeding? It was enough for the government to be alleging that the Mafia existed, why reinforce that image in the minds of the jurors with your presence? Yet here, in the new millennium, was a group of wiseguys showing up each day, looking for all the world like hopefuls at a casting call for *The Godfather, Part IV.*

It had all started in Philadelphia during the Scarfo trials in the 1980s, and continued in New York in the 1990s, when Gotti was being prosecuted. Nearly every morning during the Gotti racketeering trial in Brooklyn—the one where Sammy Gravano made his debut as a government witness—a group of half a dozen Gotti loyalists would show up, sitting in the first row behind the defense table where Gotti and his codefendant sat.

When Gotti, groomed and dressed as if he were stepping off the pages of *Esquire,* walked into the courtroom, the mobsters would stand. He would smile, wave, nod in approval. With that, they would sit back down. It was the same gesture of respect that courtroom spectators pay to the judge when he enters the room. But these guys were there for the Don. And they didn't care who knew it.

Secret society? Fuhgeddaboudit.

Other regulars at the Merlino trial each day included the wives and/or girlfriends of several of the defendants. In an amazing feat of scheduling that must have involved some fascinating conversations—"Don't come tomorrow, hon, the testimony's got nuthin' to do with me," or "Why don't you take the kids and spend a few days down the shore," or "You really shouldn't be takin' any more time off from work"— wives and girlfriends seldom showed up on the same day.

When it did happen, one of the guys—a brother, a cousin, an associate from the neighborhood—would intercept the girlfriend and either steer her back to the elevator or make sure she sat on the other side of the courtroom, well away from family, friends, and the unsuspecting wife.

Still, the issue of wife versus girlfriend, an age-old underworld conundrum, surfaced quickly. Natale, from the witness stand, made sure of that.

After being chided during cross-examination about his own extramarital affair and the money he had "borrowed" from friends to provide Ruthann Seccio with an apartment, a place at the shore, and a car, Natale lashed out.

"All the defendants have wives and girlfriends," he blurted out angrily. Then he amended that by noting that John Ciancaglini alone among them didn't cheat on his wife. Since neither Lutz nor Gambino was married, Natale seemed to be painting the defense table with a broad brush—a point that Caravasos, Gambino's lawyer, later felt compelled to point out. During her cross-examination, she noted that Gambino, who was seventy, was a confirmed bachelor who couldn't be cheating on a spouse. The jury still had to decide whether Gambino was a racketeer, but his lawyer wanted to make it clear that he was not a philanderer.

Ruthann, in an exclusive interview with the Philadelphia *Daily News,* kept the pot boiling. The tabloid had done an earlier profile in which she described her affair with Natale and her anger over his decision to cooperate. Now she gave the paper her reactions to the testimony that came up during the trial. She said that she and her family were upset over some of the defense lawyers who had grilled Natale about bickering and backstabbing among wives and girlfriends. She was particularly incensed over a defense question indicating that one of the mobster's wives considered her "trash" and that most of the wives didn't want to associate with her.

"All this time I could have trashed any one of them," she said. "Who had implants? Who's got a big mouth? Who's a whore? Who fights with their wives? Who fights with their girlfriends? Who gives bigger presents to their girlfriends than their wives?

"And what girlfriends are dating other gangsters? And who is dating her business partner?"

Ruthann acknowledged that she and Natale had had their tiffs, confirming a fight in a bar that had been the topic of some defense cross-examination.

"It was a stupid fight we had in front of them," she said, according to the article by *Daily News* reporter Kitty Caparella. "Two big-breasted girls were flaunting themselves in front of Ralph. I got pissed. He got pissed. He pushed me. I pushed him. [Afterward] he sent me flowers every day for a week."

The trial rambled on like that for weeks, testimony about murders one day, soap-opera-like accounts of petty bickering and adultery the next.

The audio- and videotape of the Joey Chang hit was riveting, the testimony of the EMT who found him dazed and bloody heart-wrenching. Joey Merlino's phone conversation with Natale, in which he described how Mikey Chang died lying next to him on a sidewalk, was chilling. But then there were the tapes of Natale on the phone from prison with his wife in the fall of 1999, mocking and belittling the same guys he had praised six years earlier as "fine young men."

They were "the macaroni mob," he told his wife, Lucia, in the call, recorded in October 1999. They were "punks," "idiots," "retards." Natale, in a comment that had Borgesi shaking his head in disgust and Merlino trying to stare right through him, boasted that he was "ten times the better man" than all the defendants and their lawyers combined.

"Their intelligence quotient, the whole group of them, can't reach double digits," he said.

The trial was a social event, a cross between a wake and a wedding. It was a place to see and be seen, to come and pay your respects, to show your support. The women wore Italian leather shoes, dress slacks, and designer tops. Or snug blue jeans, bulky sweaters, and boots. Their hair, of course, was neatly coiffed, their nails manicured and polished, their faces perfectly made up. The guys wore black leather—jackets or car coats—over black slacks, black leather shoes, and pullovers. Or they went with what some feds would mockingly refer to as "South Philly formal wear"—a designer sweat suit. The casual look usually included brand-name black or white leather sneakers, most often Nikes. Their hair dark and closely cropped, many of the younger guys featured the wet look that was in all the fashion magazines. Those who were buff—those who pumped iron on a regular basis—usually wore tight-fitting pullovers that clung to their sculpted frames.

As spring became summer and the weather changed, the jackets and leather were replaced by casual windbreakers, an occasional sports jacket, sometimes a simple open-collared shirt. Gold chains abounded.

Both local newspapers and several television stations did fashion features.

The *Daily News,* in an article entitled "Call It a Trial with Style," described the attire favored by Debby Merlino and several other mob wives as "buttery black-leather blazers" and "some of the most beautiful leather shoes ever to pass through a courthouse metal detector." The *Inquirer* countered with its own story, "Fashion Hits," noting that "most of the wives are size six or smaller" and observing that "like stylish Roman women on the Via Veneto, they invest their serious bucks in shoes (Ferragamo, Gucci) and handbags (Vuitton, Prada, Longchamp). The earrings are invariably diamond studs; the bracelets and cross necklaces, gold."

Not since John Gotti was tried in Brooklyn—and the *Newsday* coverage included both a courtroom artist and a fashion writer to report on the styles of the defendants and the spectators—had clothes been so closely chronicled in court reporting

The media were always looking for a new angle. Sometimes the testimony seemed secondary. National Public Radio did a piece on the musical tastes of the Merlino mob—using several tapes made by Previte in which music could clearly be heard in the background.

In a case that at times sounded like the script from a Grade-B gangster movie, the media had found a soundtrack. The mob favored street-corner a cappella and Motown. As Previte discussed swag, there in the background were The Shirelles and their classic lament, "Will You Love Me Tomorrow?" At another meeting, Previte was discussing bookmaking and money and there were the Skyliners doing "Pennies from Heaven."

The *Los Angeles Times,* the *Boston Globe,* and the *Baltimore Sun* all checked in with feature reports. The *Times* piece, by Stephen Braun (who had worked in Philadelphia earlier in his career), described the Philadelphia mob as "an army of mooks, a motley crew of wanton killers, traitors, braggarts, and bumblers. . . . Addicted to scheming and splintering into rival factions . . . prolific only in violence."

Even the lunch break became a topic for news reports. Within a week of opening arguments, a pattern quickly emerged. Most of the spectators, several of the lawyers, and occasionally someone from the prosecution would head to Pagano's, an Italian bistro/pizzeria/deli on the ground floor of an office complex a block from the courthouse.

Matt Pagano, the gracious and hardworking owner of the

eatery, served a lavish daily spread for the courtroom regulars. There were hot plates of pasta and seafood, a salad bar, and a deli counter featuring everything from pastrami or tuna on rye (Cutler's particular favorite) to a belly-busting chicken cutlet sandwich with broccoli rabe, sweet red peppers, and melted sharp provolone. Served in a roll that looked like half a loaf of crisp Italian bread, the sandwich was a challenge for most customers—even Fat Angelo Lutz, who usually held court each lunchtime at a table with several reporters analyzing and prognosticating.

Lutz, the only defendant out on bail, happily embraced the role of defense spokesman. He would eventually pay a price for the high-profile position he took, a position he clearly enjoyed but which also would have been almost impossible for him to avoid. As the one defendant who could leave the courthouse each day, he became a magnet for the media. He also showed up on sports talk radio in the morning and offered occasional commentary on "Mob Talk," a weekly feature that the local Fox affiliate, Philadelphia's Channel 29, began running once the trial hit full stride.

"Let's get ready to rumble," Lutz said, mimicking the well-known professional wrestling call as he arrived in court one day early in the trial. At another point he told reporters, "We don't do nothing wrong. We're just a bunch of guys who like to gamble."

Fat Angelo Lutz, who had to have a special chair brought into court so that he could sit comfortably at the defense table, began the trial making notes on a laptop computer—another special dispensation from the judge. Every day he would e-mail a synopsis of the day's proceedings to those who couldn't make it to court—until his computer privileges were revoked after he was spotted playing a video card game during a particularly dull day of testimony.

Lutz, who would take the stand in his own defense, caught a break early on when Natale said he never committed any crimes with him. "He would bring food and would leave," Natale said of Lutz's presence at mob meetings. "He was never privy to conversations. I didn't consider him a criminal."

Unfortunately for Fat Angelo, by the time the trial ended, neither the jury nor the judge saw it that way.

20

Like Ralph Natale, Ron Previte occupied the witness stand for fourteen days, and like Natale, he was subjected to a grueling cross-examination. The difference was, he seemed to enjoy the fight. Natale tried to go toe-to-toe with the defense attorneys and in the end wore himself out. Previte was a master of the sure-footed comeback, taking every punch the defense could throw, jabbing and weaving.

When Jacobs described him as "deceitful, untrustworthy, and violent," Previte agreed. But he balked when the defense lawyer accused him of playing both sides in order to save himself.

"I didn't play both sides," Previte said. "I was on one side. I was a crook."

That exchange underscored the biggest difference between the government's two star witnesses. There was no "epiphany" in Ron Previte's decision to cooperate. He didn't come over from "the dark side." He was what he was. He was just better at it than the guys sitting at the defense table.

Previte made no attempt to sugarcoat anything. There was no spin. The tapes, particularly those in which Merlino allegedly okayed the drug deals, weren't perfect. They were just the best that Previte could get under the circumstances. And despite the rolling eyes and snickers from the defense table, there were other conversations that Previte frankly admitted he had not recorded, either because he wasn't wearing a wire or because his tape had "malfunctioned." Those conversations, he said, were the ones in which Merlino clearly approved the drug deals.

Previte was prepared to live with it. He was on the stand to finish his deal with the government, not to polish it. That deal, of course—with its five hundred thousand dollars in payments and its promise of a bonus once his cooperation was concluded—also provided fodder for the defense. So did several acts of violence that had occurred while he was under contract to the FBI. Previte admitted that he'd been involved in three such incidents after he signed his 1997 deal, a deal that specifically prohibited any type of assault.

Two were altercations related to his on-again, off-again relationship with an Atlantic County woman he was dating. One involved a fight with a bouncer in the "lingerie shop" where the woman was working. The other was a confrontation with another man who was dating her.

The third incident, which occurred in municipal court in Brigantine, was quintessential Previte. "I cracked a guy for making a disparaging remark about my daughter," Previte said when Jacobs questioned him about the assault.

"You hit him?" Jacobs asked.

"I knocked him right out of his shoes," Previte said proudly. "As anyone would have done if someone had said something in an awful, foul way about your daughter."

Jacobs never asked for the details. The jury only heard about the assault. The background gives it perspective. Previte's daughter, then in her early twenties, had gotten into an automobile accident and was in court to challenge a traffic ticket. The courtroom was packed that evening. Previte had driven her there and had arranged for his own criminal defense attorney to represent her.

"We're sitting in court waiting for her case to be called," Previte had said several months earlier when he first recounted the story. "I see this guy looking at her and he's mouthing something. I can't fuckin' believe it. This guy is saying, 'I'd like to fuck you.' This motherfucker. He was a big

Italian kid. I walk right over, slide in front of him, and I come from the ground with a haymaker. Drove his lower lip right through his upper tooth. He came out of one of his shoes. I hit him so hard, the judge thought a gun had gone off."

Several court attendants and police officers rushed to grab Previte as the judge stood by hollering, "Did you get the gun? Did you get the gun?"

One of the police officers told Previte, "I gotta lock you up."

"Do what you gotta do," Previte told him.

He quickly posted bail and was released. The charges were later dropped when the victim refused to testify.

"The FBI, after they heard about it, said, 'We gotta keep these incidents down.' I said, 'Listen, that's my daughter.' And I told them if it happened again, I'd do the same thing. . . . The key in that situation is to hit first. Don't talk. The first punch, if you know how to hit, is always the best. I know how to throw a punch. This was a big kid, but he didn't know what hit him. I knocked him right out of his shoes."

From the witness stand, Previte offered the jury a grittier, street-level view of the mob, one that in some ways played into the defense strategy. Jacobs went on at length about how the FBI allowed Stanfa to continue to operate even after he had been recorded threatening to murder Merlino and Michael Ciancaglini, and even after Merlino had been wounded and Ciancaglini had been killed. The FBI, Jacobs said again and again, was more interested in making a big racketeering case against Stanfa than it was in saving the lives of any of his targets. And oh, by the way, he said, the FBI had tapes of Stanfa plotting murders and discussing them after the fact. Where were the tapes linking Merlino and his codefendants to the murders for which they were charged?

Previte, in discussing the Stanfa-Merlino war, said that Stanfa was known as "Homicide John." He also said that

Stanfa was "partially elated" after the wounding of Merlino and the murder of Michael Ciancaglini.

Why "partially," he was asked?

Because Stanfa believed, he said, "if only one guy hadda die, it shoulda been the other one."

Merlino, sitting at the defense table, showed no reaction. By that point, six weeks into what had become a marathon prosecution, talk of murder and mayhem had become mundane.

So had Joey's by then legendary penchant for stiffing those who lent him money. Previte testified that he'd once given Joey a fifteen-thousand-dollar loan, cash Merlino needed to pay back an earlier loan from Natale—money Joey had blown on a trip to Vegas. Previte used money from the FBI to pay Joey; Merlino paid back about two grand, then stopped making payments.

"If Joey had known the FBI was fronting the money," one of the defense lawyers later quipped, "he wouldn't have paid back any of it."

Previte also discussed the stolen property deals, the Rolex watch he had given Merlino, the car he leased—again with FBI funds—and the phones he provided. The only time he balked about a request for money from Merlino, he said, was when Joey tried to hit him up for fifteen hundred dollars for one of the Christmas parties for the homeless.

"I thought it was a scam," Previte said. "I thought everything he did was a scam."

Previte, of course, had one thing going for him that Natale didn't. He had the tapes he had made. The conversations involving the swag were devastating, incontrovertible. The drug tapes, those recorded in Boston at least, were everything he said they were.

While Previte talked about his entire career and provided a detailed account of his life wearing a wire, he spent most of his time on cross-examination fending off Jacobs's attempts to undermine the evidence of cocaine deals that the government said Merlino had approved. Jacobs highlighted one tape in which Merlino cautioned Previte against dealing drugs in Boston.

"Don't even get involved with that," he said. "Seriously, Ron, it's bad."

That, coupled with Borgesi's flat-out prohibition, gave the defense a clear opportunity to argue that the Merlino organization opposed drug dealing. Jacobs then tried to argue that Merlino was discussing stolen property, not cocaine, when he talked about deals in Boston with Previte.

Previte, of course, was having none of it.

"Georgie nixed it always, from the beginning to the end," Previte said. "Georgie never wanted any part of it. Joey did. . . . For the right amount of money, he would do it. . . . There was money to be made. That's what Joey was all about. Money . . . The only thing that turned into cash in Boston . . . was cocaine."

On a Sunday afternoon in the middle of his second week on the witness stand, Ron Previte went to dinner with his girlfriend and her three children at the Margate Pub. They had decided to take a ride to the shore that afternoon. Despite the fact that his name was in the headlines every day and his picture was showing up in the newspapers and on television, Previte had continued to live as normal a life as he thought possible.

As they walked through the bar toward the dining area, Previte noticed a guy staring at him. "He was tall, heavyset, had a moustache. About fifty years old."

Previte wasn't sure if he knew the guy or not. When Previte got up to go to the men's room, he walked past the guy and nodded hello. The guy ignored him. Later, he walked over to the table where Previte was working on a plate of clams and spaghetti.

"I know who you are," the guy said. "You're Ron from Hammonton. I don't like what you're doing." Previte's girlfriend, no shrinking violet, called the guy a "disrespectful piece of shit" as he walked back to the bar. Previte said at that point he was going to let it go.

They finished dinner.

"She, naturally, had the lobster. Most expensive thing on the menu. The kids ate. We had dessert, some coffee, whatever. I get up to go to the men's room again before we leave. As I walk by, this guy goes, 'Fuck you, you fat motherfucker.'"

The guy then took a swing. Previte slipped the punch, grabbed the guy in a headlock, picked a bottle of Heineken off the bar, and slammed it into the guy's head. Then he hit him again.

"He goes down. There's blood and glass all over the place."

Previte went to the bathroom, returned to his table, paid the bill, and left.

The trial was in recess the next day when he got a call from John Terry, the FBI agent who had succeeded Jim Maher as head of the Organized Crime Unit. "Have a good weekend?" Terry asked.

"Yeah," Previte said.

"Did you go out to dinner?"

"Yeah, had a nice dinner."

"Did you have dinner at the Margate Pub?"

Previte laughed. He knew Terry knew. "I hit a guy with a bottle," he said. "He had it coming."

In fact, police later told Previte's lawyer, who filed a report about the incident, that the guy at the bar said he was at fault, that he had started the altercation.

Previte, in the middle of testifying against the mob, was still living life on his terms.

"I ain't running," he said. "And I ain't gonna let nobody fuck with me. I signed on to do a job and I did it. I was out there taking chances, putting my life on the line. . . . If I walk outside this building and somebody else says something to me, I'll hit him with a bottle too. That's who I am, and that's not changing."

*   *   *

In all, there were sixty-four days of witness testimony during the Merlino trial. The first witness took the stand on March 30, 2001. The last appeared on July 5. Twenty-eight of those days—more than forty percent of the trial—were taken up with the testimony of Natale and Previte. The government built its case—at least its most serious charges—around their testimony.

Support came from Scafidi and Caprio, who also testified about mob murders and life in the underworld. But there were enough minor points of discrepancy in their testimony to give the defense at least some wiggle room.

Scafidi, at times near tears, came across as one of the most genuine witnesses, although to this day, on the corners where Merlino associates still hang, he is described as a "lying rat" who turned on his friends.

Tommy Horsehead detailed the murders and attempted murders he was involved in, and recounted the personal and financial betrayals that marked his time as a Merlino associate.

"Joey was no good all his life," he said. "He kept robbing people. . . . That's why we're here."

But in many ways, Scafidi said, Stanfa was no better. He "put this crew of misfits together," Scafidi said. "And I was one of them."

Caprio, one of the last witnesses called, backed up Natale's story about the murder of Sodano. But there were other tapes, made by Casale, that allowed the defense at least to imply that the murder was part of a robbery, or part of an internal, Newark-based mob dispute.

Pete the Crumb also brought some levity to the courtroom, describing how he had been involved in organized crime literally all his life. The seventy-year-old wiseguy told the jury that when he was an infant in a baby carriage in Newark, his father, a bookmaker, used to push the coach around the neighborhood collecting bets and making payoffs.

When he was just eight years old, Caprio added, he and some friends were pinched by the cops after they were caught breaking into a warehouse and trying to steal comic books. They were released, he said, after one of the arresting officers, who was from the neighborhood, asked them to sing an Italian folk song.

Jacobs, always ready to use whatever the prosecution gave him, came back to that story when he was cross-examining Caprio.

"Is that when you learned you could get out of jail if you sang the song the police wanted to hear?" he asked.

Neither Philly Fay Casale nor Bobby Luisi was called as a witness.

Casale's agreement with the government had started to unravel after Caprio agreed to cooperate. In his original offer to the feds, Casale had admitted to his role in two murders prior to the Sodano hit. Caprio tied him to four, a discrepancy prosecutors decided they'd rather not put in front of a jury. That, coupled with Casale's conviction for child molestation, made him a less-than-attractive witness.

Boston Bob Luisi was another matter. Luisi and his three associates had all pleaded guilty to the cocaine-trafficking charges. In fact, the tapes of that FBI sting—an almost textbook example of how to run an undercover operation—were played for the jury in the Merlino trial when the undercover agent, Mike McGowan, took the stand.

Jacobs then used those tapes to try to undermine the case against Merlino. In Boston, he said, it was clear what was going on. Here were the audio- and videotapes of the deals being negotiated, the drugs being delivered, and the money being paid. Where in Philadelphia were there any tapes like that tying Merlino to cocaine trafficking?

Maybe if Luisi had taken the stand to corroborate Previte's account of the conversations with Merlino, the case would have come in stronger. But up in Boston, Bobby had found Jesus. He had shaved his head and begun walking around the protective custody wing of the jail where he was being held, carrying a Bible and quoting Scripture.

He was telling people—his family, his lawyer, any friends who still talked to him—that he'd been born again—that he was talking to God, and, more important, God was talking back. He heard the Lord's voice. Luisi said he was planning to write a book. He already had the title—*From Capo to Christian.*

All of that made its way to Philadelphia. In the middle of the trial, word started to spread that Luisi might not be taking the witness stand after all. While prosecutors refused to comment, those in the defense camp privately and happily fueled the rumor. Luisi had gone south on the feds. They were afraid to use him, afraid to put him on the stand because they didn't know what he might say.

Other witnesses who were called included FBI agents, Philadelphia and Pennsylvania State Police detectives, and extortion victims. One of the best and most effective was Michael Casolaro, who reluctantly recounted how he was abandoned by John Ciancaglini and shaken down by the Merlino mob.

Casolaro clearly did not want to be on the stand, which only made his testimony more compelling.

Before the trial ended, two defendants got up. First Angelo Lutz, then John Ciancaglini.

Lutz's street-corner style, however, didn't play well. Instead of entertaining, he came across as brash. There was not enough humility and entirely too much hubris. It was not one of Fat Angelo's better performances. And it would cost him dearly.

Ciancaglini also blew a chance, however slight, to separate himself from the other defendants and score some points with the jury. Other than the Casolaro testimony, the tape of his conversation with Previte was the most devastating piece of evidence against him. It tied him directly to the mob, made it clear that he knew what was going on, and showed that he was a willing participant—albeit one smart enough to make himself "scarce" whenever he could.

But instead of trying to spin that evidence in his favor, Johnny Chang denied the obvious. Instead of trying to use his past—his conviction in the 1980s, and the devastation that had befallen his family—to his advantage, he was caught lying about it.

The only logical, noncriminal explanation for the conversation with Previte was for Johnny Chang to admit that those were his words but to claim that he was only saying what he knew Previte wanted to hear. He knew that, because he had been through it all before. He had been involved with Scarfo. He had done time. So had his father. But now one brother was dead and the other was a cripple and he, Johnny Chang, had learned his lesson. That was the one shot he had with the jury.

Instead, Johnny Chang claimed that he was never involved with the mob—that he was set up by Previte.

The prosecution quickly came up with a photo taken in the 1980s of Johnny on a boat in Fort Lauderdale with Nicky Scarfo and a dozen other wiseguys from that era. This was the Scarfo that Johnny Chang claimed he didn't know, the mob that he was never a part of.

For both Lutz and Ciancaglini, beating the rap by taking the stand was a long shot at best. Neither, however, did anything to help his chances.

Closing arguments lasted five days; then, on July 14, 2001, the jury got the case. Deliberations took the better part of a week. The crowds were gone now. Only the family members—wives, brothers, sisters, mothers—showed up on a regular basis. Those who couldn't make it had cell phones or beepers at the ready. There was a calling system in place if and when the jury came back. It would take about twenty minutes to get from South Philadelphia to Sixth and Market Streets, the site of the federal courthouse.

The jury had several questions during deliberations. One would have an impact that is still being felt by Merlino.

There were thirty-six counts in the indictment. The two primary counts were charges of racketeering and racketeering conspiracy. All seven defendants were charged with those crimes. Each racketeering charge included a subset of crimes—so-called predicate acts—used to support the racketeering allegation. Different defendants were charged with different predicate acts: the murders, the attempted murders, the gambling, the extortion, and the receipt of stolen property. The predicate acts against Merlino also included the Sodano murder and cocaine trafficking.

By law the jury had to find that at least two of the predicate acts had been proven in order to find a defendant guilty of racketeering. There were more than thirty predicate acts listed in the indictment.

In addition, the case included thirty-four other specific counts; charges ranging from the murders of Billy Veasey and Anthony Turra and the attempts to murder Joe Ciancaglini Jr. and Anthony Milicia to the receipt of stolen property, bookmaking, and extortion. Almost all the predicate acts were repeated as separate counts—all except the Sodano murder.

In addition to finding the predicate acts either proven or not proven, the jury was required to vote guilty or not guilty on each separate count. Those were the ground rules as the deliberations began. Along the way, they got muddled.

In the midst of deliberations, the jurors sent out a question: If they decided that two of the predicate acts against a defendant had been proven—in other words, if they had already found the defendant guilty of racketeering—could they list any predicate acts on which they had not reached a unanimous decision as not proven?

Over the objections of the prosecution, Judge Hutton told the jury it could. Prosecutors had argued that any predicate act for which the jury could not reach a unanimous decision should be designated undecided; that is, neither proven nor not proven. An appellate court would eventually rule that the prosecution was right; that the judge's ruling was incorrect.

Shortly before 5 P.M. on Friday, July 21, the jury sent out word that it had reached a decision. By 6 P.M. the courtroom was jammed. The jurors filed in, and, after they had taken their seats, the forewoman stood and began reading from the lengthy verdict sheet.

Merlino was first. He and Jacobs stood as the woman, a school librarian from Philadelphia, prepared to announce the findings.

"Guilty," she said, when asked how the jury had found for the first charge in the indictment, racketeering conspiracy. An anguished sigh went up in the courtroom. Wives and mothers tightly held hands. But then the forewoman, responding to questions posed by the judge's clerk, began to detail the findings for each predicate act.

"Conspiracy to murder . . . John Stanfa?"

"Not proven," she said.

"Attempted murder of Joseph Ciancaglini Jr.?"

"Not proven."

"Conspiracy to murder William Veasey?"

"Not proven."

"Murder of William Veasey?"

"Not proven."

It would take more than an hour for the forewoman to read all the verdicts, but the pattern held for Merlino and the other defendants. Merlino had been named in sixteen counts of the indictment. He was found guilty of eight: racketeering, racketeering conspiracy, extortion, bookmaking, three counts of receiving stolen property, and one count of conspiring to receive stolen property. But he was found not guilty of the Billy Veasey, Anthony Turra, and Anthony Milicia shootings; not guilty of two counts of cocaine trafficking; not guilty of two stolen property charges; and not guilty of one extortion charge.

All the other defendants were also convicted of racketeering and racketeering conspiracy, but as in the Merlino verdict, the decisions were a mixed bag of "proven" and "not proven," "guilty" and "not guilty."

But for each defendant charged with an act of violence—Merlino, Borgesi, Ciancaglini, Mazzone, Angelina, and Gambino—the verdict was the same: "not proven" as a predicate act, "not guilty" as a separate count charge. The jury rejected every act of violence in the government's case.

Family members began to cry, then hug one another, when they realized what was happening.

The verdict was a rejection of everything Natale had brought to the table. To a lesser degree, it was also a rejection of Scafidi, Caprio, and, at least in terms of the cocaine charge, Previte.

"In a nutshell, we won," Eddie Jacobs said as he was surrounded by a jubilant crowd outside the federal courthouse later that Friday night.

"The message the jury sent to the government tonight," said Jack McMahon, Angelina's lawyer, "is that you better come to the table with some evidence and not just the words of people like Ralph Natale. Words are cheap."

Defense attorneys happily—though in most cases incorrectly—predicted prison terms of seven years or less for their clients. "This was the Volkswagen of racketeering charges," Jacobs said of the guilty verdicts. "We face the lowest possible [sentencing] levels."

Lutz, caught up in the moment, seemed to welcome the prospect of prison for what he then expected would be a relatively short term. "Guess what?" the four-hundred-pound young mob associate said. "I'm gonna lose some weight now. As long as they got a job for me in the kitchen and they got balsamic vinegar, I'm gonna be all right."

Acting U.S. Attorney Michael Levy, at a press conference, bristled at the suggestion that the jury's verdicts signaled a "stunning defeat" for the government, as some television and radio reports were already calling it. "Any time I see a defendant facing seven years in jail I think it's stunning," Levy said. "But *defeat* is not a term I would use with it. This jury clearly rejected their vision of themselves as just small-time thieves and found that they were, in fact, La Cosa Nostra."

But this wasn't the La Cosa Nostra of Angelo Bruno or Carlo Gambino. Or even, for that matter, of Nicky Scarfo and John Gotti. This was the La Cosa Nostra of the new millennium, an organization devoid of honor and loyalty. Built around fear and treachery, this was an organization that self-destructed, that turned on itself, first in scams and flimflams designed to generate cash, then in murderous and wanton acts of violence, and finally in turncoat testimony from the witness stand that may or may not have been based on fact.

The prosecution had presented Ralph Natale as Don Corleone, the wise and wily Godfather who knew it all. The defense portrayed him as Uncle Junior, the out-of-touch puppet mob boss of *The Sopranos*.

It was left to one of the jurors, who asked to remain anonymous, to put it in perspective.

"We couldn't convict somebody on the hearsay of Ralph Natale," he said in an interview two days after the verdict was announced. "Natale could have stayed home or in jail or wherever he came from. He was not a factor."

EPILOGUE

On the night the verdicts came in, Ron Previte was on his way to dinner with his girlfriend. He first heard about the convictions on the car radio, then watched the news reports on TV from the bar in the restaurant.

"All the television stations were carrying the story," he said. "We were sitting at the bar in the Old Mill Tavern, this place I like in Mays Landing. They got great pork chops there. That's what I had that night. I'm sitting there eating my pork chops. It was just a night out for me. I wasn't elated and I wasn't disappointed. I thought they could have been convicted of everything. But for me it was just a J-O-B. It was a job. Almost everything I gave them came to fruition and I guess I took some pride in that. Whatever I do, I do it right. That's how I've always operated. But there was nothing to celebrate. It wasn't about seeing them guys in jail. They didn't scare me. I didn't need for them to be in jail. . . . The convictions were just the end of the job. It was what it was."

The convictions, of course, were more than that.

They signaled the end of the brief but highly chronicled Joey Merlino era, a period of South Philadelphia underworld history unlike any other. And if Philadelphia was, in fact, a microcosm of the American Mafia, the convictions were more nails in the coffin of La Cosa Nostra.

The final nails? Probably not. The mob, the Mafia, La Cosa Nostra, will always exist in some form. No longer monolithic and not nearly as sophisticated as some of the new and emerging crime groups—the Russians and Eastern Europeans, for example—the Mafia will always have a place in the American underworld. But its stature was fading long before Merlino took over the top spot in Philadelphia.

That he and Natale could wield power—however briefly and however ineptly—was the real indictment of the organization.

With the seven defendants in jail awaiting sentencing—Angelo Lutz's bail was revoked shortly after he was convicted—things grew quiet on the streets of South Philadelphia. Joe Ligambi headed up what was left of the organization, but the crime family membership chart—which in the Bruno era had boasted seventy members and during Scarfo's heyday included about fifty—was down to about a dozen.

The feds, as always, were working another mob racketeering case. Testimony from the trial had indicated that both the FBI and the Pennsylvania State Police had already targeted gaming and loan-sharking operations linked to Ligambi, including a lucrative video poker machine distribution network. Pete the Crumb had also implicated Ligambi—however tentatively—in the murder of Ron Turchi, a mob hit still under investigation.

Roger Vella, sitting in a prison cell facing a first-degree murder charge for the brutal 1995 slaying of Ralph Mazzuca, had decided the life of a wiseguy was not for him. In September 2001 he cut a deal with the Philadelphia District Attorney's Office and the feds, agreeing to cooperate in ongoing investigations into organized crime.

In exchange, Vella was allowed to plead guilty to third-degree murder for the Mazzuca slaying. He faced a maximum of twenty years, but hoped for substantially less. At his plea hearing, Assistant District Attorney Edward Cameron said Vella had implicated "others" in the murder. Vella eventually pleaded guilty to a federal racketeering charge as well, admitting to having played roles in both the Billy Veasey and the Ron Turchi murders. He also confessed to drug trafficking, extortion, and to attempting to intimidate witnesses on behalf of the mob.

Despite the feds' insistence that the racketeering convictions were a victory, it was no secret that the FBI and federal prosecutors wanted to put Merlino and those around him in jail for murder. They hoped Vella could help them do that.

Merlino was named in another indictment one month after the verdicts came down in the racketeering case. The U.S. Attorney's Office in Newark announced that Skinny Joey had been charged with murder for the 1996 slaying of Joe Sodano. The feds argued that since the Sodano murder was only a predicate act in the racketeering case, they could file the charge again as a homicide count in a separate indictment. The predicate act argument and the faulty instruction by Judge Hutton were two prongs of the legal position prosecutors used to justify recharging Merlino.

Because Hutton had incorrectly instructed the jury, prosecutors said, there was no way to determine if the "not proven" the jury had listed after the Sodano murder on the verdict sheet was a unanimous decision.

Chris Warren, who had represented Lutz at the trial, took over Merlino's case in Newark. Claiming double jeopardy, he argued before a Third Circuit Court of Appeals that the new indictment was, in fact, the same charge for which Merlino had been acquitted. "Not proven," he contended, was the equivalent of "not guilty," and court error notwithstanding, the government should not be permitted to try Merlino twice for the same crime.

The appellate court ruled two-to-one in favor of the prosecution. The U.S. Supreme Court declined to hear the case. As this is being written, prosecutors are preparing to retry Merlino for the Sodano murder. They will rely, in all probability, on the same witnesses, Natale and Caprio, and the same evidence that was used in the racketeering case. If that's not double jeopardy, defense attorneys contend, then it is a glaring example of a major flaw in the RICO Act that the government has used so effectively over the past fifteen years to dismantle the mob.

Judge Hutton imposed sentence on the seven defendants in the racketeering case over an eight-day period early in December 2001. Each defendant had his own hearing. While prosecutors had privately complained that Hutton was too lenient and allowed defense attorneys too much leeway during the trial, they had no cause for concern during the sentencing process.

Hutton hammered the mob. Every defendant was given a term at or near the top of his sentencing guideline. Merlino and Borgesi each got fourteen years. Mazzone, Ciancaglini, and Lutz each got nine years. Angelina got six and a half years. And Gambino got seventy-one months.

Lutz got more time than two made members of the organization. The nine-year sentence was, many observers believed, Judge Hutton's response to the antics of the flamboyant and outspoken mob associate. The media attention that Fat Angelo had generated came back to haunt him. It may have carried more weight with Hutton than Natale's flat-out assertion that he didn't consider Lutz a criminal, but merely the guy who brought the food.

The sentencing hearings allowed the government one final word on each defendant. Assistant U.S. Attorney David Fritchey, who wrote most of the sentencing memos, didn't miss his chance to send Joey and the others off with a flourish.

Lutz, Fritchey wrote, "clings to his association with the Philadelphia La Cosa Nostra family like a drowning man clings to a piece of flotsam in the middle of the ocean. . . . His sense of self-worth is tied desperately to his association with Borgesi, Merlino, and his other unworthy masters in the mob."

Borgesi, in Fritchey's memo, was described as a punk and bully. "One three-minute tape tells the hearer everything he needs to know about what George Borgesi is about and, for that matter, about what the Philadelphia La Cosa Nostra family is all about," Fritchey wrote, referring to the infamous Lutz beating tape.

"In this one tape [Borgesi] exposes himself for what he is—a cruel, petty, greedy, boastful bully . . . a coward and a loser down to the very bottom of his empty soul. It is for moments of pathetic domination such as these that bullies who never grew up aspire to become members of organized crime."

Fritchey heaped similar disdain on Mazzone, Angelina, Ciancaglini, and Gambino. But he unloaded on Merlino. "Joseph Merlino grew up with the aspiration of becoming a gangster," the prosecutor wrote. "He was a mob associate while still a teenager and he became a full-fledged convicted felon by the age of twenty-two. Nothing that has happened in the intervening seventeen years has changed the direction of his life. . . . He has accomplished exactly what he sought to do. . . . At the age of thirty he became a made member of the Philadelphia La Cosa Nostra. . . . He did not join La Cosa Nostra to become a decent person. He joined to seize the opportunity to make more money and enjoy personal power . . . that membership in organized crime provides. His crimes are deliberate, premeditated and systematic. . . . He will never be a decent, law-abiding, legitimately productive citizen. He has committed his life to the values of La Cosa Nostra and is the confirmed enemy of a civilized society based upon the rule of law."

In the same week that Merlino was sentenced, agents with the Pennsylvania Attorney General's Drug Task Force raided his home in South Philadelphia. Joey was sentenced on December 3. Agents hit his house on December 6.

Skinny Joey's wife, Deborah, answered the door that morning in her pajamas. It was a little before 7 A.M. In the house were the Merlinos' two young daughters, a woman who worked as the children's nanny, and, upstairs hiding under a bed in his undershorts, the man the agents had come looking for.

Billy Rinick, a brash young South Philadelphia cocaine dealer and suspected murderer, was the guy found under the bed. He would later claim that he was a "friend" of both Joey and Debby and that he was just looking out for her.

In fact, agents working a drug case against Rinick had seen him in Debby Merlino's company on several occasions. She was even picked up on one surveillance tape dropping him off in South Philadelphia where he was about to complete a cocaine deal.

There has never been any evidence or suggestion that Deborah Merlino was involved in Rinick's drug dealing. Authorities, however, said he was a frequent visitor to Merlino's home during the racketeering trial. And they clearly enjoyed putting him there in his underwear hiding under a bed just three days after Skinny Joey was sentenced to fourteen years in jail.

And still the soap opera continued.

Three weeks later, on December 27, 2001, Bobby Luisi went into court in Boston and withdrew his guilty plea in the drug case. His cooperating agreement in tatters, Boston Bob decided to go with Jesus, pitching his born-again conversion and government entrapment as a defense. That's what he offered the jury when he went to trial a year later. The jury wasn't buying, not after the prosecution buried Luisi under an avalanche of audio- and videotapes made by Previte and Mike McGowan.

Boston Bob, who had bought his way into the Philadelphia mob, may have paid the steepest price of all. He was sentenced to twenty years in jail, ten more than his three Boston associates had gotten for pleading guilty and more than any defendant in the Philadelphia case.

Assistant U.S. Attorney Ernie DiNisco, at Luisi's sentencing, described his brief attempt to cooperate as "half-baked" and said the government had not derived any significant information from the wannabe wiseguy. Luisi, who showed up for sentencing wearing a velveteen yarmulke emblazoned with the Star of David, said that the government had dumped him as a witness because of his religious conversion and his refusal to lie.

"Born again mobster gets time to pray—in prison," read a headline in the *Boston Herald* the next day.

Luisi's conviction made it a clean sweep for Previte. Everyone who had been targeted after he put on a wire was either convicted or had pleaded guilty—Natale and the five codefendants in the meth case, Luisi and his three confederates in the cocaine ring in Boston, and Merlino and the ten defendants named in the racketeering indictment.

Twenty-one mobsters in jail and a mob family shattered.

The government, over the years and including a bonus after the trial had ended, paid Previte about one million dollars for his work.

Was it worth it?

First, consider that Previte put his life on the line not only each day when he put on a wire, but every day that he met with or provided information to law enforcement. How much should society be willing to pay for that? Then consider the market. Consider, especially, the deal Ralph Natale got.

Natale began cooperating *after* he was indicted. A mob boss who was looking at life in prison, he hopes to soon walk out of jail. What did he give the government?

None of the murder or attempted murder charges he brought to the Merlino prosecution stuck. Blame it on the judge's error, on faulty prosecution, on poor investigative techniques, or on good defense work. Maybe a combination of all four. But the bottom line is that the government convicted Merlino and his codefendants on the charges it had in place *before* it made a deal with Natale. In reality, he added nothing to the case.

Natale also testified against Camden Mayor Milton Milan. Milan was convicted of fourteen counts. Only three involved organized crime. The mayor was going to jail even without Natale's testimony.

In May 2003, Natale was called as a witness in the bribery and corruption trial of Danny Daidone and James Mathes, the former Camden City Council president who got a four-thousand-dollar diamond ring for his girlfriend by promising to help Natale get city contracts. The trial was a rehash of testimony heard in the Milan and Merlino cases, but it did include a disclosure for the first time from the FBI that Natale had first broached the idea of cooperating in 1998, after he was jailed for his parole violation.

Months before his supposed "epiphany," Natale was trying to work a deal. He had offered, according to FBI Agent John Terry, to wear a wire and record conversations with Merlino and others if the feds would spring him from jail and guarantee him a minimum jail sentence for his crimes. The feds nixed the deal, Terry said, because they were not permitted to promise any kind of sentence.

On July 8, 2003, Daidone and Mathes—a mob gofer and a politician who couldn't or wouldn't deliver the goods—were convicted of conspiracy and wire fraud charges. They are the only two defendants whose convictions rest primarily on the testimony and credibility of Ralph Natale.

And for that Natale will get a get-out-of-jail card.

It says here the government got a hell of a lot more for its money and society got a much bigger bang for its buck when the feds cut their deal with Ron Previte than either party did when the FBI fell in love with the idea of flipping Ralph Natale.

Ron Previte was sentenced on August 7, 2003, by Judge Stephen M. Orlofsky, the same federal court judge who had presided over the DeLaurentis trial. As expected, Previte got no jail time. He was given five years' probation for his role in the Hammonton extortion that had sent DeLaurentis to prison. He faced no other charges.

Prosecutors and FBI agents turned out in support. Mary Futcher, the assistant U.S. attorney who prosecuted the Milan and Daidone cases, called Previte's work "unprecedented." He put his life on the line, she told Orlofsky, and in the process had helped make cases that brought down three mob bosses—John Stanfa, Ralph Natale, Joey Merlino—and dozens of their associates.

Asked to comment before the sentence was imposed, Previte told the judge that he realized his life "had not been exemplary," but he hoped his work for the FBI had made up for some of the things he had done.

A week later, sitting in a restaurant in South Jersey working on a plate of shrimp and pasta, the still massive former wiseguy talked about what he had done and frankly admitted he was puzzled about what might come next.

Hiding in plain sight, he had shown up at the popular Italian eatery casually dressed and wearing a black Adidas baseball cap on his now clean-shaven head. He was satisfied, he said, with the sentence and with the two-hundred-and-fifty-thousand-dollar bonus payment that had pushed his earnings over the one-million-dollar mark. But like a corporate executive who had been given a golden parachute and then downsized out of a job, he no longer had anything to do. And after living on the edge for the better part of forty years, Ron Previte said he was bored with the leisure life a new identity and relocation brought him.

"I'm used to doing things," he said, between forkfuls of

pasta. "All my life I worked. Usually I worked at crime, but I worked. I can't sit around. It makes me nuts."

It was typical Previte. Philosophical, self-deprecating, but always surprising, he had told his story in bits and pieces during dozens of similar interviews over a six-year period, adding anecdotes and asides that helped if not explain, then at least accurately describe who he was.

During one discussion about music several months earlier, he said that he had always been a big Elvis Presley fan; Presley's "Welcome to My World" was one of his favorite songs. He talked about how he sometimes found himself humming the tune when he was setting up a scam or about to confront a sucker who owed him money or who was the target of a shakedown.

"I think I'd like that played at my funeral," he said with a grin.

But he could also be introspective.

During another lunch he told a story about his grandmother, about how she had been a midwife in Hammonton after coming to this country from Italy, about how the wiseguys back in those days would always come to her if they got one of their girlfriends pregnant. She would be forced, he said, to perform abortions.

He remembered how he once saw his grandmother sitting in the backyard sobbing. He was just a boy at the time and asked his mother what was wrong.

"She said, 'Grandmom's sad because she hears all the babies crying,'" Previte said quietly.

For almost four decades Ron Previte roamed the underworld, making deals, busting heads, setting his own agenda. When he was on top of his game, the six-foot, three hundred pound wiseguy moved easily in gangland circles that stretched from Philadelphia to Boston, from New York to Atlantic City.

He was a player, a major moneymaker, an intimidator and

occasional head buster. Operating with the same guile, wit and stone-cold bravado that he displayed in the underworld, he then became the FBI's secret weapon in its war against the mob.

"Fortune favors the bold," he said when he first started talking about his life in the underworld. Unknowingly he was quoting one of the tenets of a Renaissance philosopher he had never read, but whom he understood intuitively.

Ron Previte was an underworld Machiavelli, and his decision to turn on Stanfa, Merlino, Natale, and the others was a classic gambit right out of *The Prince*.

His story, the saga of an underworld mercenary, provides a definitive and easily understood explanation for how and why the American Mafia has come undone. In the end, Previte was the mob's worst nightmare. He was a wiseguy wearing a wire. But even before he reached that point, Big Ron was an underworld wild card, opinionated and outspoken, fearless and fiercely independent. He was a mobster whose only allegiance was to the individual he saw looking back at him in the mirror each morning.

None of that has changed.

As he looks back on it now, Previte realizes that he came along about thirty years too late. His mentality, his demeanor, and his attitude were more suited to the mob of the 1950s and 1960s, the glory days of the organization, when, he says, "real gangsters" ran the families. Of course, back then, a Natale or a Merlino would never have risen to the top of an organized crime family. But then in those days an ex-cop like Previte would never have gotten close enough to a mob boss to record his conversations.

"There's no honor anymore," Previte says.

Now sixty and still living within an hour's drive of his old haunts, Ron Previte has opted not to go into the Witness Protection Program. He has a new government-provided identity, but his appearance is largely unchanged. He says he will continue to live life on his own terms and in his own fashion.

"Men flourish when their behavior suits the times and fail when they are out of step," Machiavelli wrote five centuries ago.

Previte has never been out of step.

He used to bristle at the "Fat Rat" moniker that his former associates slapped on him and that the media quickly adopted. Now he laughs.

"Where are they?" he says of the defendants who mocked him.

"And where am I?"

They, of course, are in jail.

And he?

He is wherever he wants to be.

ACKNOWLEDGMENTS

Ron Previte called me about eight years ago after I had written several newspaper articles about his ties to Philadelphia mob boss John Stanfa. He said I seemed to have an undue interest in his business, and he wanted to have lunch. We've been talking ever since. And while it would be several more years before I learned he was cooperating with law enforcement, his insights, philosophy, and general take on life in the underworld were invaluable to me as a reporter. This book could not have been written without his candor and willingness to speak in detail about what he did and what he was thinking while he was doing it. For that, I thank him.

There are also dozens of others—underworld figures, law enforcement investigators, prosecutors, and defense attorneys—who have always been willing to share their thoughts. Some may not agree with the tone and tenor of this book, but I appreciate their input.

I would also like to acknowledge several editors at the *Philadelphia Inquirer* who helped shape the telling of this story as it was occurring. These include Julie Busby, David Taylor, Mark Wagenveld, Francisco Delgado, Conrad Grove, Michael Mills, Deirdre Childress, and Avery Rome.

Finally, a special thank-you to Judith Regan for giving me the chance to write this book, and to Cal Morgan, whose editing made it better.